Battles with Model Soldiers

Battles with Model Soldiers

Donald Featherstone

Line illustrations by W. R. F. Jenkinson

Drake Publishers : New York

ISBN 87749 044 9

Drake Publishers, Ltd
381 Park Avenue South
New York
New York 10016

Printed in Great Britain

Contents

List of Illustrations

PART ONE
Table-top Battles

1
What is Wargaming?

THE SHATTERED village showed no signs of life; the main street
lay empty but for a brewed-up Sherman tank nestling amid
heaps of rubble and charred timbers. Suddenly a shell screamed
through the air, bursting high on the tower of the already
ruined church and seemingly to act as an alarm bringing the
area to life. Cautious, semi-concealed movements caught the eye
—the long barrel of a 75-mm gun slowly poked round the wall
of a cottage followed by the snub nose of the Panther tank in
which it was mounted; a crouching file of grey-clad German
infantrymen moved slowly forward, hugging the walls of the
battered houses that lined the littered village street. Suddenly
the tank commander spotted the cleverly concealed anti-tank
gun and madly traversed his turret on to the target . . . too late!
The anti-tank gun fired first and the German tank shuddered to
a standstill, smoke rising ominously from it. Overlooking the
battlefield, the German commander realised with despair that he
had just lost his last armoured vehicle. The British commander
surveyed the field with satisfaction, issued crisp orders that sent
three Sherman tanks crawling from behind a low ridge. Slowly
he smiled, realising that he had the battle in his grasp!

The German commander swallowed, his voice choked as he
uttered the fateful words: 'I concede the battle!' Then, as the
two generals strolled off to supper, he began to puzzle out where
he had gone wrong. For, happily, this was not real war, no men
were being killed and no damage was incurred except to a few
cardboard houses, plastic tanks and plastic soldiers—who do
not leave behind them plastic widows and children!

No, this was a 'war-game' played on a table-top six feet square,

with realistic terrain that included roads, rivers and a village; hills were graduated slopes of Plasticine and vegetation made of lichen-moss. The tanks were plastic vehicles made by Airfix and assembled, converted and painted by the opposing 'generals'; the troops likewise were low-priced 00-gauge Airfix modern figures—but the pleasure the pastime gave to both commanders was beyond price!

Table-top battles with model soldiers have been fought for centuries as a pastime and as a means of military training. It is a hobby making such great forward strides that, in time, it may well rival the already well-established hobbies of model railways and miniature-car racing. At some time or other in their lives, most boys collect model soldiers but only a small proportion stay with the hobby as they grow older. This perceptive minority have passed over a threshold that leads to many hours of pleasure and satisfaction as they apply skill and research to turn out colourful and realistic model soldiers and to use them intelligently in a highly competitive and fascinating hobby rather than allowing them to gather dust on a shelf. Because there comes a time when a collector will turn a reflective eye upon these glistening parades, conscious that they are wasted in their glass cases, and consider if there is not some practical way in which they can be put to better use. To line them up and fire toy guns at them would be far too damaging, besides seeming rather childish (he is already a little touchy about letting his friends know he plays with 'toy soldiers'!). The competitive spirit that lies latent within us all combines with Man's innate militancy to suggest using the soldiers in their true role—to fight wars, but in miniature.

Some very well-known, even famous, men have found enjoyment and relaxation in battling with model soldiers. Sir Winston Churchill fought large-scale battles with his brother; Robert Louis Stevenson and H. G. Wells formulated complicated rules and fought battles over large areas of floor and garden. In the latter days of the nineteenth century H. G. Wells wrote the first popular book on wargaming, he called it *Little Wars* and described it as 'a game for boys from 12 years of age to 150 and for

the more intelligent sort of girl who likes boy's games and books'.

Well-known people of this day and age collect model soldiers. Charles Heston, screen portrayer of Ben Hur and El Cid, has a large collection; so have actors Peter Cushing and Douglas Fairbanks. Television actor Deryck Guyler is an authority on the Roman Army and has a very fine array of this period with which he fights battles. The Swedish Ambassador in London, and at least two famous dress designers, are collectors but this does not mean that it is a hobby only for men of wealth and position. A start can be made by a small cash outlay and a wonderful collection acquired by expending time more than money.

With some accuracy, the pastime has been labelled 'chess with a thousand pieces' and a battle can be fought with as many soldiers on either side as desired. Each of these small figures is a man with a little metal or plastic personality of his own. Sometimes he fights individually and sometimes he fights as part of a regiment, but always he plays a manful part with his qualities (and faults) reflecting the character and temperament of the man who moves him. Campaigns and battles can be planned ahead, prior movements being made on maps until two opposing forces come into contact. Then the scenery is laid out on the table-top battlefield, made realistic by hills, villages, trees, rivers, roads and bridges on which the small soldiers move bravely forward. There are no cowards in wargames, and every warrior lives to fight another day. Naval and air warfare can also be re-created on the table-top; there are books which fully describe this facet of warfare and offer suggested rules for the manoeuvring of miniature ships and aircraft.

Battles with model soldiers are neither a solace for frustrated generals nor a means whereby downtrodden sergeants can work off their spleen, rather are they a skilful relaxation, capable of being made as complicated and far-reaching as desired, or so simple that even young children can take part.

Probably because of the host of coloured uniforms and the vast amount of information that can be so easily obtained, the

Napoleonic Wars are the most popular period in which to campaign. The Wars of the Ancients provide another fascinating field—the heraldry and armour combine with chariots and war elephants, knights and archers to present a unique spectacle, as well as affording a realistic impression of how men fought in days gone by.

When we were children, most of us received as birthday and Christmas presents the large 54-mm figures which are still obtainable in toyshops. Soon we realised that they were really far too big to allow of a successful game unless one had unlimited space in which to battle. Today most wargamers use the small 00-gauge plastic figures which have recently appeared on the market, supplementing them with more expensive metal figures of the same size. Using these readily available and suitably scaled models on a table of reasonable size, battles can be fought which allow of tactical manoeuvring, so producing realistic and highly enjoyable actions.

Apart from the actual battling with model soldiers, there is also the great enjoyment to be obtained from making one's own model soldiers. First, the 'master' figure has to be built up, then the mould is made from one of the synthetic-heat resisting rubbers. Next, the molten metal is poured in, then the gleaming silver casting is extracted. After cleaning off the 'flash' that sometimes attaches to the casting, the figure is painted to conform to the information on uniforms obtained from books, plates and prints.

It is not within the scope of this volume to explain how to make model soldiers—but it will be obvious that the hobby does not end with campaigns and battles. Many pleasant hours may be spent painting model soldiers, turning out scenery, and carrying out research into battles, campaigns and uniforms or tactics of the period. Times passes pleasantly when browsing in museums, bookshops and galleries, and the hobbyist will find that he can never know too much and that there is always more of interest around the corner. Campaigns can be chronicled with drawings and photographs to form nostalgic reminders of past battles. Correspondence inevitably results when fellow wargamers begin to contact each other, exchanging notes and giving information in the friendliest and most helpful fashion. In fact, it is sometimes difficult not to let the 'branch-lines' sidetrack one from the main business of battling with model soldiers.

Once counted in tens, wargamers now number thousands, many gathering regularly in clubs where battles of all periods are fought and members participate in annual inter-club contests at national conventions. In addition to the ever increasing list of books dealing with the hobby, there is a monthly magazine, *Wargamer's Newsletter*, now in its tenth year of continuous publication. Wargames, in brief, can be great fun, not only for their realistic re-creation of battles in any chosen historical period, far removed from the horrors of the real thing, but also for the opportunity they provide to lead a winning army in which there are no mutinies, no disobedience to orders and no lagging under fire!

2
The Three 'S's' of Battling with Model Soldiers

MANY WARGAMERS are first attracted to the hobby by being pressed into service as an opponent to a friend who has already been bitten by the bug. After that, it is not difficult to learn how to begin collecting model soldiers and fight campaigns with them. Other would-be wargamers may not be so fortunate and before launching out on their own they would be well advised to study the 'Three S's' of campaigning with model soldiers.

1. SOLDIERS
2. SCENERY
3. SITUATIONS

SOLDIERS. In the 1890s, makers in France, Germany and Great Britain were turning out the large 54-mm (2½-in) model soldiers that most men remember from their boyhood days. They were made of metal and easily broken. Nevertheless many of these remarkable model soldiers made by William Britains of London are still in existence and are avidly sought after by collectors who carefully repaint them. These were the figures used by H. G. Wells in his immortal 'Battle of Hook's Farm', described so vividly by word and picture in his book *Little Wars*. Indeed, they are the type of soldier which, prior to the advent of Airfix 00-gauge figures, were used in wargames by the majority of beginners. In the United States, a number of collectors have gathered together in a nostalgic group called the Wellsian Society. They collect Britains' 54-mm figures and fight wargames with them to the precise rules laid down by H. G. Wells.

Page 17   (*above*) A miniature 'scena' with a British tank of World War I vintage crushing forward over a German machine-gun position; (*below*) a 'Stand of Pikes', the 18-ft pikes would have been thrust belligerently forward and the musketeers would have been sheltering beneath their outstretched security

Page 18   (*above*) A French Foreign Legion fort under attack by a group of Bedouin Arabs; (*below*) British infantry, supported by a light tank and an armoured car, flush-out hillmen on the North West Frontier of India—*circa* late 1930s

In order to fight table-top battles, one must obviously have two armies of model soldiers, and the simplest way in which the novice wargamer can begin is to amass two relatively small forces of these 00-gauge plastic figures. They have much to recommend them. They are easily transportable and armies can be taken to fight 'away' battles without having to be carefully packed in cotton wool to save them from damage. Unlike the larger 54-mm models, they can be used in considerable numbers over small battlefields, thus providing an interesting game in which flanking movements and widespread manoeuvres are possible. Even the smallest village shop now sells these colourful packages of excellently moulded and accurately scaled plastic figures covering a wide range of types and periods. The range available at any one time varies; as an indication, at present the following are issued by the leading manufacturer, Airfix.

Guards' Colour Party. In full dress with bearskin, this set includes an ensign with a standard, an officer with a sword and a party of marching guardsmen, together with sentries and sentry-boxes.

Guards' Band. Again in full dress, this is a 44-piece marching band with a drum-major and bandsmen.

British Infantry Combat Group. Contains officers, NCOs, machine-gunners, men in various fighting positions and a stretcher party.

German Infantry Group. The World War II enemy to match the preceding set. Included are officers, grenade throwers and an anti-tank gun with crew.

British Eighth Army Group. Wearing shorts and steel helmets, these men are positioned as machine-gunners, riflemen, mine detectors etc.

Afrika Korps. Providing the opposition to the Eighth Army, this set of Rommel's men includes anti-tank guns, senior staff officers and troops in various fighting positions.

United States Marines. This set contains men in various combat positions and includes bazookas, flame throwers and an assault boat.

Japanese Infantry. The opposition to the American Marines, this set includes officers brandishing swords, buglers and infantrymen with rifles, sub-machine guns and grenades.

British Paratroops. In addition to men in various fighting positions, this set includes collapsed parachutes and supply containers, together with a mortar and crew.

Russian Infantry. Completing the main combatants of World War II, this set includes a machine-gun, a mortar, officers, NCOs, and infantry in action poses.

World War I British Infantry. Wearing the flat-topped cap of 1914–15, these 'Tommies' include signallers, a wiring party with a trench mortar and crew.

World I German Infantry. Again, the early period of the war, these men are wearing pickelhaubes and include a field officer, two infantry officers and a heavy machine-gun with crew, plus riflemen.

World War I French Infantry. Set in a later period of the war, these men are wearing French steel helmets and the set includes cyclists, a signaller with pigeons, a bugler, a standard-bearer and a variety of infantrymen in action poses.

World War I American Infantry. A 48-piece set including men in fighting positions, together with light machine-guns.

Foreign Legion. A 48-piece set which includes riflemen in various positions, a bugler, plus mounted and foot officers.

Bedouin Arab Tribesmen. The foe to fight the Legion, this set includes camels and riders, Arab horsemen and tribesmen on foot armed with rifles.

American Civil War Union Infantry. These Federals include officers, buglers and riflemen in various action poses.

American Civil War and Confederate Infantry. Includes officers, bugler and riflemen in five positions—all moulded in appropriate Confederate grey.

United States Cavalry. Although garbed more for the later Indian Wars, this 36-piece set can readily be used in the Civil War and includes an officer, bugler, standard-bearer and troopers with a selection of horses.

American Civil War Artillery. Made so as to be capable of being used either as Union or Confederates, this 33-piece set contains a mounted officer, horses, limbers and guns with commanders and crews.

Robin Hood and his Merry Men. In this 40-piece set one can find Robin Hood, Friar Tuck, Little John, Maid Marion, and men fighting with bows and staves.

Sheriff of Nottingham. Robin Hood's enemy, this set includes mounted and foot knights, bowmen and swordsmen.

(Both these sets are ideal for conversion into medieval armies, such as those which fought at Agincourt.)

Romans. Equipped as at the time of Julius Caesar, this set contains legionairies, a chariot with horse and driver, a centurion and a standard-bearer.

Ancient Britons. A foe to the Romans, this set contains chariots, chieftains and numerous figures with swords, shields, spears and slings etc.

Red Indians. This 42-piece set includes galloping and standing horses, chieftains, squaws and a wide selection of braves.

Cowboys. Includes cowboys on foot and horseback, some armed with revolvers and some with rifles.

Waggon Train. This is the third of the Wild West sets and models the famous waggon train of the new frontier. The set includes a snap-together nine-part waggon assembly, with household goods and families.

Tarzan. This set includes Tarzan himself, some animals and natives with spears and shields. With a little painting and conversion, these men make excellent tribesmen, Zulus, etc.

Waterloo Highland Infantry. This set consists of forty-five figures in twelve different positions, including a mounted officer, a piper, a drummer, a colour-bearer and men fighting in six different poses.

French Cavalry (Cuirassiers). Each box contains sufficient riders and horses to make up eleven cavalrymen in a most colourful and attractive uniform.

Obviously, these latter represent the first of a considerable future range of Napoleonic figures, and their arrival should further enhance the already great popularity of those inexpensive plastic wargames figures.

Many collectors will start off with these figures and continue happily with them, while others may wish to branch out into some period not made in this range and seek out the cast metal figures which, though considerably more expensive, offer a greater variety of choice. This is simply because collectors rarely stay in the same groove and most of those who campaign seriously with model soldiers appear to incline more towards the heavier and more stable metal figures, even though plastic figures can hardly be beaten for their variety, definition and detail. Wide ranges of wargames figures in 00 and other scales can be obtained from a number of makers throughout the world and those collectors and campaigners who wish to purchase metal figures should find that most of their requirements can be satisfied by the makers listed in Appendix 3.

In addition, there are numerous Continental makers of the 'flat' figures; a two-dimensional figure in 20-mm or 30-mm scale, stamped out of sheet metal. These figures are immensely popular in Europe because of their availability in *any* period and in *any* position at relatively low prices. The best-known makers are also shown in Appendix 3.

Although the plastic figures obviously do not cover every type of soldier, period, position and style likely to be required by the discriminating collector, not the least valuable of their assets is the ease with which they can be converted into soldiers of vastly different types and periods. Such conversions offer an intriguing and limitless scope for ingenuity and skill—with the minimum of tools and equipment, the most complicated 'surgery' can be performed to turn a figure into something completely different. The simplest conversion consists of replacing heads and limbs, putting a different top to a trunk and removing plastic from or adding it to the figure. Bending limbs into different positions will also considerably alter a figure; the figures should be scalded in boiling water and then quickly bent into position before being held under a cold running tap to set the new posture. Head or limbs can be satisfactorily joined by pushing the blunt end of a small cut-down pin into the figure component and then pushing on the head or limb with a layer of adhesive between the two parts. Astonishing things can also be done by the use of a hot knife-point or, better still, a small soldering iron.

A mediocre conversion may be transformed into a very creditable one by competent painting, and here plastic figures can be given a convenient handle by sticking a long pin into the underside of the base. After painting, the pin is stuck into a block of Plasticine and the figure left to dry. Before painting, always wash the figures in a detergent solution to rid them of grease, then spray them with matt-white undercoat and paint as many figures as possible at one time. Not only is this the quickest way but it economises on paint, as a brushful can be completely used. Matt paint gives a more realistic appearance and dries more rapidly than gloss, but is perhaps more vulnerable to handling. Many collectors spray the completed figure with a gloss varnish to enhance its appearance and give it a protective coating.

Of all the processes involved in military modelling, painting is perhaps the most difficult, yet it is also the most interesting and rewarding part of the procedure. A simple basic model can be transformed into a work of art by a high standard of paintwork but the wargamer has first to decide whether he wishes to assemble easily-recognisable armies with a minimum of time and trouble, or to be not only a wargamer but also a collector and have armies in which each man is a little work of art. The former relies on appearance in mass, being rewarded when his troops are laid out on the table-top, whilst the latter has troops that may be individually gloated over! It is entirely a matter of preference and in either case, as with most other things in life, what you get out of it depends very much upon what you put in.

Before beginning, it has to be decided whether the soldiers are to have a flat matt finish or a possibly unnatural gloss. The experienced model maker, of course, wants neither—he prefers those parts of his soldier which should shine (such as boots, weapons, etc) to be in gloss, but for the clothing and parts that are normally flat to have a matt finish.

The beginner will probably wish to paint his soldiers as quickly as possible. There is a quick-fire method of painting wargames figures and the easiest way if one wishes to go into old age with reasonable eyesight is to fix them by their bases on to strips of wood with perhaps an inch between each figure and say, twenty figures to a strip. The strip is held in the left hand, the brush is loaded with the required colour and the painter proceeds happily down the line painting all the light blue trousers or all the brown packs or all the pink faces and so on.

There are about ten different colours to the average soldier, but one or two of which predominate. So pick one of these main colours, say, for example, a red tunic, and with a big brush apply red to the whole of the top part of the figure. Do not worry about covering up the head, the pack or other parts of the figure because each new colour will be carefully painted along the line where it joins with another colour.

Thus, when painting a British Guardsman with a red tunic and dark blue trousers, the red will probably slop over part of his trousers, but the basic blue colour will cover the uneven red line. Be very careful about this joining point or edging. Having applied the two main colours to the soldier, next put on some of the smaller colours, the black of the boots or pack, the red stripe down the side of the trousers and the black of the collar. The stand, or base, of the figure is usually painted last of all—paint it a bright green as this helps to bring out the uniform colours. Paint the face and hands next to last. If the figure is being held in the hand singly and not put on a stick, paint the musket the last thing before the face. This means that the musket is available to hold on to whilst painting; by doing the flesh colours after the musket, it can be covered where the hands held it without having to repaint it, even if paint slopped over it when painting the musket. The final detailed painting of wargames figures, such as belt, straps and epaulettes, are the tedious parts of the job. Well done, they turn a mediocre figure into a good one. These fine lines can sometimes be put on with a mapping pen and Indian ink.

Through experience, painters of model soldiers have learnt a number of short cuts, such as the use of a magnifying glass when painting to show up smaller details and so enable them to be accurately depicted. Fix the magnifying glass so that it is standing erect, hold the stick with figures attached by the left hand on the other side of the glass and peer through whilst painting with the right hand.

There are many varieties of paints on the market which are eminently suitable for painting model soldiers. For easy work, use a quick-drying paint so that it is possible to get on quickly with the rest of the figure. Purists will prefer to use oil colours, which give a fine end-result but take a couple of days to dry. Remember, the sooner you get those figures painted the sooner you can start fighting battles with them.

SCENERY. Reasonably satisfactory battles could be fought on a table-top without any sort of scenery whatsoever, the two armies forming up on the bare expanse, advancing, making contact and fighting to a finish. But battles are not fought on perfectly plain drill squares lacking the shelter and tactical advantages of hills, valleys and trees. To fight a battle without any scenic features is like staging a play against a single-colour backdrop—it can be done but it leaves a lot to the imagination. When attractively set out to represent countryside, a table-top battlefield adds stimulus to the forthcoming contest, as well as realism by providing tactical factors such as cover from fire and concealment. Scenic features on the wargames table also provide objectives by means of which it can be decided who has won and at what stage, because warfare is largely a matter of one side losing or gaining some advantageous topographical position. Thus, when reconstructing an actual battle, it is essential that the scenery should resemble the field in real life—Waterloo will require the ridge, Hugomont, La Belle Alliance and La Haye Sainte.

Many wargamers believe that realistic scenery is equally as important as the uniforms or formations of the men who are to fight over it. Beginning with simple battles on the kitchen- or dining-room table, they go to immense trouble to make a much larger surface of blockboard or chipboard, or hardboard supported by wooden battens, erected on trestles in a special wargames room. Some of these battlefields measure 8 ft × 5 ft or

9 ft × 6 ft, or even cover as much as 72 square feet. However, it must always be remembered that the table should not be wider than six feet because that is the maximum distance which can comfortably be stretched when moving soldiers in the middle of the table. The battlefield can be as long as desired, the length making possible vast-ranging battles with exciting flanking movements.

Obviously, the wargamer with a special wargames room has an edge on the man who has to put up and take down his scenic effects on the same evening, but the latter can assemble very realistic scenery providing he has storage space available. It is necessary for him to lay his bare scenic base on the table-top in sections and then make up prefabricated scenery on squares of hardboard or insulation board, conveniently sized to fit in with each other. For example, on a scenic table 8 ft × 5 ft, his squares will need to be multiples of 8 in—thus 8 in, 16 in and 24 in. On one square a hill can be built, on another a farm or village, whilst a third may have a sunken road section or a river and a marsh. Each square fits closely alongside another so as to form a road running across the table, passing over a hill, crossing a river first by bridge and then by a ford. At one point there may be a side road, leading to a farm, which winds through a wood and skirts a ploughed field. This 'scenic square' system is ideal providing storage space is available to pack the squares away when not in use. These squares can be assembled, jig-saw fashion on the table, to give a most realistic scenic effect and they can be endlessly permutated to build up different battle terrains.

White polystyrene is the ideal material to use in the construction of scenic sections that are cheap, light, clean and easily produced. It can be carved and shaped, sandpapered and cut—the best tool to use is a serrated-edged breadknife, although the sharp blades of handicraft knives are also suitable. Weighing very little, polystyrene tiles are ideal for use in the construction of hills. Using a large irregularly shaped piece for the base, shaped smaller pieces can be piled on to it and glued into position. When dry, it can be sandpapered or carved into the desired shape. Roadways, caves, paths and shell-holes can be cut into it and its large flat areas meet an essential requirement of all scenic effects for campaigns with model soldiers—that the soldiers themselves must be able to stand up on it.

Only special adhesives are suitable for use with polystyrene, the best being Copydex. It will be found, too, that some paints act as solvents and melt the polystyrene to a sort of 'candy floss'. Mix the paints well with thinners as the polystyrene is absorbent and requires paint with a good flow quality. Poster or watercolour paints can be used but a great deal of them is required and they usually give a rather pastel-shade effect. Polystyrene, when cut or torn or roughed with a file, makes excellent rough-cast walls for buildings, bridges, etc, especially if painted grey and perhaps scattered with some kind of scenic powder before the paint is dry.

Most wargamers have leanings towards sand tables, perhaps because professional soldiers are trained in tactical operations on very large sand tables in most of the world's military establishments. Only the wargamer lucky enough to have a permanent set-up can, however, go in for this ultimate refinement, which has to be built up on a strong, rigid base-board with built-up sides about 6 in high. Obviously, because sand is heavy, the table on which the base-board is placed must be of the stoutest construction. The sand itself can either be silver sand, which is fine and clean, or some less expensive material such as plasterer's sand. Before working the sand into the necessary undulating scenic effects it needs to be dampened for ease of moulding, which means that the shallow-sided table must be lined with a waterproof or plastic material.

The sand can be moulded into hills, ridges, river beds, or sunken roads by using a plasterer's trowel and a child's seaside spade. Using cement colouring powders mixed with water, the sand can be painted with a large brush or sprayed from a tin full of the colouring material that runs out through holes pierced in its bottom. Ploughed land can be painted brown and cornfields yellow, roads are painted in with yellow ochre and excellent rivers are made by pouring a very thin green/blue mixture into the prepared watercourse so that it flows and forms its own river bed.

Although sand tables look wonderful and provide dead ground in which bodies of troops can be hidden out of range, their preparation takes very much longer than any other method of making up scenic effects. Sand also has the unfortunate habit of creeping over the sides of the board and lying grittily on the floor below, while unfortunate 20-mm soldiers have been known to vanish in its depths, to emerge surprisingly many games later. Perhaps every wargamer should at some stage or other have a sand table; whether or not he will persevere with it is a matter of conjecture.

The scenery over which campaigns with model soldiers are fought varies immensely according to the temperament, attitude and available time of the wargamer responsible for making up the terrain. Few people will dispute that there is much more enjoyment to be obtained from manoeuvring troops over a realistic battlefield than one composed of crudely chalked roads and rivers, or with bridges and houses made of thinly-disguised matchboxes. Nevertheless, realistic scenic effects may not be within the reach of every wargamer, especially when there is a problem of storage, since these scenic effects, being bulky and fragile, cannot easily be 'stacked' away.

The solution to such a problem may be found in an assortment of scenic items recently introduced by Merbelen Limited and known as Bellona Battlefields. Light in weight and strongly made in earth-coloured PVC that can be packed away in a small area, these 'instant battlegrounds' are inexpensive—even the largest and most ambitious piece so far marketed can be purchased for the price of four boxes of plastic soldiers. They can be cut and used in combination with each other so as to form elaborate and ambitious set-ups, they can be mounted on chipboard and blended into dioramas two or three feet square and they can be coloured with Humbrol, or poster paints, in a very short time.

At the time of writing the majority of the available pieces fit best into a modern setting, although the Redan is a classic de-

fended earthwork which can be used for any period from Marl-
borough onwards. Measuring $6\frac{1}{2}$ in × $10\frac{1}{2}$ in, it is a hastily im-
provised fortress made out of sandbags and timber with row
upon row of wicker gabions. Other pieces are a 1914–18 trench
system with firing platforms and revetted sides, and a piece
known as the Menin Road, which is a ruined landscape with a
removable cellar roof which also forms the ground for a wrecked
building. There are gun positions, fighter dispersal bases, gun
and mortar positions, trenches and slit trenches, sand-bagged
emplacements, Japanese bunkers, ruined buildings, pill-boxes,
encampments, defence works, tank traps and lengths of wall and
river section.

This compact, self-contained 'instant' scenery saves hours of
setting up before a battle and taking down afterwards. It also
enables the wargamer to concentrate on painting his soldiers
rather than attempting to make a large-scale piece of scenery
which, if he lacks ability, may bear only a coincidental resemblance to what was intended. Again, most wargamers, particularly in their novice days, have a passion for a battlefield
crowded with scenic effects, so that their soldiers are frequently
bogged down and a dull battle results. Using one or two choice
pieces of Bellona material, positioned tactically on the battlefield with a few trees and smaller items to add colour, results in a
far less crowded and more realistic battlefield on which it is
possible to move one's troops with freedom and wage a much
more enjoyable and effective battle.

There is no reason why campaigns with model soldiers should
not be fought on battlefields that look like miniature replicas of
the real thing. The model-railway enthusiast has the highest
standards of scenic effects and wargamers should seek to emu-
late him. Making up scenic effects to act as a backcloth for
campaigns with model soldiers can be just as fascinating as
assembling the armies themselves.

SITUATIONS. When scenery has been laid out and model
soldiers placed upon it the resulting situation must be fought out
to rules which require the action to bear a reasonable resemblance to the tactics and methods of the period in question. Any-

one who has been campaigning with model soldiers for any
length of time will acknowledge that wargames rules reflect the
character and temperament of their devisor, so that the steady,
plodding type of man will hold his hands high in horror at the
dashing tactics permitted by the rules formulated by a man who
goes through life in a free and easy fashion. This is why few
wargamers have any patience with rules other than those they
have worked out for themselves! There is no end to the realism
that rules can bring to campaigning with model soldiers because
they can cover such varied factors as the firing of different types
of weapons, ammunition supplies, wounded and dead soldiers,
prisoners, fatigue, morale, and such like. In other words, rules
can be either very simple or very complicated.

Certain basic factors must, however, be common to all rules.
For example, it is feasible to assume that neither infantry nor
cavalry will move as fast when climbing hills as they do on flat
country, therefore hills should be climbed at, perhaps, half-
rate. Similarly, since troops do not move as quickly across
country as they do on roads, increased distances should be
added to their normal move. Men in the shelter of a house or
behind a stone wall are not as vulnerable to fire as men out in
the open—the rules must allow for this. Rivers can be classified
into those which are uncrossable except by bridges or fords, and
those which take perhaps one complete move to struggle across.
Or there may be mere streams or creeks, rating perhaps deduc-
tion of a third of the distance from the move. Walls and hedges
cannot be gaily surmounted and taken in the stride, so a portion
of the distance must be deducted from a soldier's move to compensate for the time and effort he has expended in climbing
these obstacles. These are but a few of the common sense factors which must be taken into consideration when formulating
any set of rules.

Taken at its simplest basic level, a battle with model soldiers
is conducted in a series of moves, during each of which all one's
troops can be moved, those within range can fire their weapons
and, if the situation calls for it, parties of men may engage in
hand-to-hand fighting (known as a mêlée). In its most elemen-

tary form, a wargame begins when each general lines up his army in any desired formation along the base-line at the rear of the table. The scenery will have been laid, with its hills, roads, rivers, villages, woods, etc, so that each general can have some idea of his tactical plan. Moving at the specified rates, each army travels across the table until contact is made. Moves are taken alternately, in that a dice is thrown or a coin tossed and the winner has the choice of moving first or second. This may be done before each move, or the original sequence may be continued throughout the battle. If side 'A' moves first in the first turn of the game then side 'B' moves first in the second turn, side 'A' moves first in the third turn and so on. Deciding before each turn which side has first or second move introduces the element of luck or frustration to be found in real warfare. The man who moves second has the privilege of firing off his weapons before his opponent. Straight away, we arrive at one of the most compelling aspects of wargaming—the need to make a decision. Often it is beneficial to move one's troops first, but then your opponent has the right to fire upon them before you can fire back. An alternative method is for both sides to move simultaneously, each side starting its move from the left, or a dice may be thrown at the beginning of each move to decide whether moving takes place from the right or the left of the table. Again, this brings a certain amount of realism to the battle but has the built-in disadvantage of each general moving his troops while frantically watching his opponent moving his troops at the other end of the table. Thus each will hastily carry out his movements so that he can get to the other end of the table to cope with the threat to that flank. Moves may be timed, such as when each side is allowed five minutes to move everything. As this is frequently an impossibility, some unfortunate units do not get moved at all and are left stuck out in the open awaiting the onslaught of the greatly superior enemy forces facing them.

There are also many variations to the practice of starting each game from opposing base-lines. For example, a defensive battle may be fought where one man is allowed to lay his troops down on his entire half of the table and the other man, who is the attacker, brings his troops in from his own base-line. The attacker is then able to see the dispositions of the enemy but at the same time the enemy has a little time in which to alter his dispositions to meet the apparent threat. Another variation is to erect some sort of curtain or barrier across the middle of the table so that each general can lay his troops out without his opponent being able to see his dispositions. Many horrible shocks await each general when the curtain is lifted! The commander of Blue army discovers to his horror that the enemy has a great preponderance of cavalry on their left flank whereas he has but one puny infantry regiment facing them. His own strong left flank, with which he intended to battle away at the enemy, are massed threateningly enough, but unfortunately there is nothing opposite them. This makes for hasty rearrangement and even the best-laid plans of battle can come to naught when the curtain is lifted.

If it is assumed that a 'wargames day' takes eight moves and night takes four moves, then it is possible to calculate the time of day at any period during the battle. For example, a battle may begin at midday when the contact between the two forces was made, and it will have to die down at nightfall, often providing the losing side with a good excuse to get out. Rather than just fighting a game until there are five men left on one side and two on the other, it is better to have some sort of objective— perhaps to agree at the beginning that side 'A' must try to hold the village, or that side 'B' is attempting to cut 'A's' line of communication or take possession of a vital cross-roads or river bridge.

A further variation is to erect the curtain across the middle of the battlefield and then lay down one's forces right up to the curtain so that when it is withdrawn the two sides are in immediate opposition. From all of which the potential wargamer will have realised that the variations are innumerable and that rules can be formulated to cope with them all.

Both sides having moved, they may then fire. After firing, any mêlées that have come into contact are fought out, and when they are concluded that move is over and the next begins in exactly the same manner.

Few wargamers actually fire missiles at their figures, preferring not to expose their skilfully painted soldiers to the highly detrimental effect of pieces of wooden or metal rod hurled at them with considerable velocity. Fire-power is represented instead by throwing dice; one dice is thrown for every five men, and this is called a volley. At maximum musketry range 3 is deducted from each dice, at half-range 2 is deducted and at close range only 1 is deducted. Thus, ten men firing at the enemy at maximum range will throw two dice and score perhaps a 6 and a 4; deduct 3 from each dice thus giving a total of four enemy dead. Guns can be fired in a number of ways—by measuring the distance from the muzzle to the objective and then throwing a dice. At maximum range, the dice has to come up 6 to register a hit, at half-range a 5 or 6, and at close range a 4, 5 or 6. Once a hit is registered, another dice is thrown and the total scored on it is the number of enemy killed. An alternative method is to have a circle of transparent plastic or a large curtain ring—known as a 'burst-circle'—and, when a hit is registered, to place the centre of this circle over the point of the hit. All the enemy covered by the burst-circle are deemed to have been killed.

Hand-to-hand fighting takes place when two opposing bodies of troops come into direct contact and form a mêlée. Here, infantry count as one point each and cavalry as two points each; one dice is thrown for every five points involved in the mêlée. Thus, a force of five cavalry would count as ten points against ten infantrymen also totalling ten points, so that each side would throw two dice and the totals killed would be HALF the dice score. For example, if the five cavalrymen threw 3 and 1 they would kill two men, but if the infantry threw 4 and 4 they would kill four points, or two cavalrymen. To introduce a note of realism, the shock impact that a force of cavalry would inflict when charging into a mass of infantry is represented by adding 1 to each of the dice thrown by the cavalry.

When the hand-to-hand fighting is completed it is necessary to seek a means of ending the mêlée, since no group of men would carry on fighting interminably. This is done by assessing the 'morale' of the two opposing forces, the side with the lowest morale rating having to retreat. Let us say that three cavalrymen remain on one side, while the enemy has seven infantrymen left. Each side throws a dice and multiplies the points value of their remaining soldiers by the number thrown on the dice. Thus, the three cavalrymen (equalling six points) throw a 5 giving them a total of thirty points and the seven infantrymen (equalling seven points) throw a 4 which gives them twenty-eight points. The infantry's morale rating being the lower of the two, they would then have to retreat a normal move-distance.

The constant use of dice may give the impression that campaigns with model soldiers are a sort of 'snakes and ladders' affair. This is incorrect, because the fluctuations and luck of the dice-throw simulate the ebb and flow of fortune on the battlefield. In real life, if ten men pointed their rifles at the enemy they would not necessarily hit ten of the enemy; so with the dice, they can either hit the maximum or only a few.

The method of assessing infantry at one point and cavalry at two points in the case of mêlées has another use in that it enables wargamers to fight battles with armies which are equal in strength but different in composition. For example, Red army may have 100 infantrymen making a total of 100 points. Blue army may have twenty-five cavalry (making a total of fifty points) and fifty infantry to give them their overall total of 100 points. Similarly, points may be assessed for guns (suggest ten points per gun) and for weapons and forces of varying periods.

This brings us to the recognition that different rules are required for different periods of history. Although tactics in all periods bear a basic similarity (the outflanking movement by a force of cavalry is the same operation as an outflanking movement by a squadron of tanks), the type of warfare fought by Henry V and his archers against the French at Agincourt differs greatly from the manner in which the Federal army beat the Confederates at Gettysburg in 1863, 600 years later. In its turn, the American Civil War and its tactics are immeasurably different from those used in World War II. It is no use attempt-

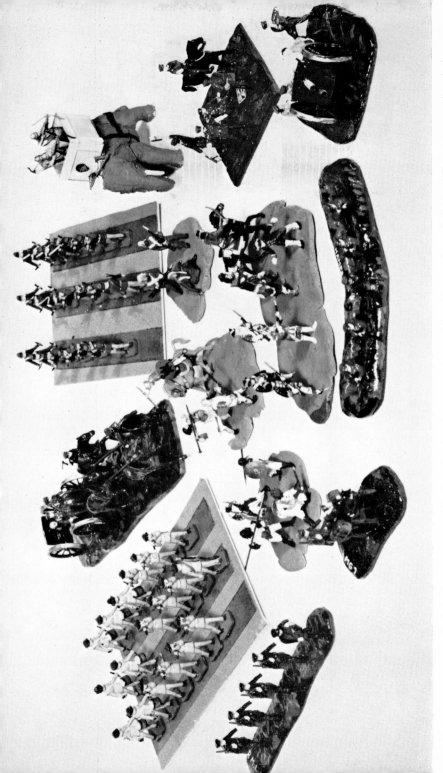

Page 35 Groups of Airfix figures, some made up and painted, others converted into figures of a completely different type and period

Page 36 Federal and Confederate forces moving towards each other during Game No 3; the situation at the end of the first move

ing to campaign with the Roman Legion under the same rules used to represent the fighting between the French and the British during the Napoleonic Wars, and there is an even wider gulf between those rules and others for a modern battle involving tanks and automatic weapons.

Nevertheless, there is a basic similarity of tactics running throughout military history from the very beginnings of time until the present day. Soldiers have always manoeuvred, they have always fired or thrown missiles at each other, and they have fought in hand-to-hand combat, although progressively less frequently since the advent of gunpowder. When it comes to campaigning with model soldiers, these three factors can be simplified into what might be called the 'three M's of wargaming'.

1. Movement
2. Missiles
3. Mêlées

MOVEMENT: Just as in real war, movement (manoeuvring) on the wargames table is primarily directed at obtaining a superiority of force at a crucial point or, by threatening the enemy's flanks, to compel him to retreat or force his surrender by surrounding him. Perhaps the American Civil War cavalry leader, Bedford Forrest, best summed up the situation when he said that his principle of war was '. . . to get there fastest with the mostest'.

With the relatively small numbers of troops used on a wargames table they can often be so diminished by losses that there are not enough left to attempt any such grandiose manoeuvres as outflanking or surrounding. Even so, it is possible to obtain a local superiority of numbers on the wargames table by means of a careful movement of one's troops. For example, by using a form of the 'Oblique Order' of Frederick the Great (see the section in Chapter 9 dealing with this period) and employing a majority of troops on one advancing flank whilst the other smaller flank is withdrawn (refused), it is possible to establish a numerical superiority which, at the very least, can disconcert your opponent.

Opportunities for outflanking an entire force come rarely on the wargames table, although single regiments or part-formations can sometimes be caught in this unenviable position during the heat of battle. Beyond the scope of this work, but fully described and illustrated in books of a more advanced nature, is the manner in which large-scale outflanking movements can be carried out with the aid of maps. Before transferring to the wargames tables, both generals carry out their initial movements on maps covering an area perhaps three or four times larger than the actual wargames table. Moving at specified rates, each general attempts to force a contact with the enemy in which he will have the advantage of a superiority in numbers and of being in a position to outflank his opponent. Once the contact has been made, the scenery of that particular area is reproduced on the wargames table and battle commences.

The relationship of movement on the wargames table to ground-scale is a facet of the hobby which gives rise to considerable argument, and it is possible to become bogged-down and frustrated by a host of relatively unimportant details if the wargamer delves too deeply into this rather complicated aspect of wargaming. If it is to be acknowledged that the game is the thing, to be played for enjoyment and relaxation and not as a form of military exercise or training, then the campaigner with model soldiers will be well advised to confine himself to a simple set of rules with a reasonable resemblance to the period in which he is fighting. However, if he has the sort of mind that likes to explore far afield and look into all the corners, he will probably, at some time or other in his wargaming career, find himself embroiled in abstruse calculations that seemingly do little to enhance enjoyment of the game. Certainly the author, in more than fifteen years of wargaming, has never consciously considered other than the simplest aspects of movement and the ground-scale, yet has consistently enjoyed in every period of military history campaigns which have provided realistic resemblance to the tactics and manner of fighting of those periods.

Practical experience has shown that the movement distances specified in a set of rules for the various arms involved (ie, infantry, cavalry, guns, etc) has an all-important bearing on the manner in which the wargame is carried out. Small move-distances of only a few inches tend to make the battle progress slowly and haltingly, the opposing armies take far too long to come into contact and precious time is wasted in preliminary moves with far too little actual fighting. On the other hand, if the move-distances are lavish, rates of feet rather than inches, then a wild, hell-for-leather game results, with frantic cavalry sweeps across the table in which units are contacted and outflanked in an unrealistic manner. Something between the two is required and many years of experience indicate that the following suggested move-distances give a reasonably fast-flowing game with a realistic outcome.

Infantry: 12 in per move.

Cavalry: 18 in per move.

Guns (horse-drawn): 18 in per move, allowing 3 in for un-limbering *and* limbering-up.

These move-distances are suitable for any wargames table upwards of 5 ft × 4 ft in size. The three demonstration battles described later were all fought on a table 8 ft × 5 ft. Obviously, the distances can be scaled down to suit the proportions of a smaller table.

It has already been explained how to organise 'stepped-up situations' using minimum numbers of figures. Now let us continue to the logical conclusion by fighting-out these battles move-by-move, so that the reader can see for himself exactly how easy it all is and how smoothly they ran.

Move-distances are known already, and the ranges of the small arms and artillery must next be considered. The infantry regiments concerned in these affrays are composed of twenty men plus an officer and a standard-bearer. One volley of rifle fire (represented by one dice) is permitted for every five men. Thus a full strength regiment of twenty men will fire four volleys by throwing four dice. If the regiment has only fifteen men remaining, then three volleys are fired and it is usual to allow eight men, for example, to fire two volleys, whilst seven men should fire only one. The maximum rifle range that will fit in with the move-distances is 24 in and firing is carried out as follows.

At 24 in (long range)—deduct 3 from each dice.

At 12 in (medium range)—deduct 2 from each dice.

At 6 in (close range)—deduct 1 from each dice.

When the target consists of men sheltering behind a wall or in a house, deduct 1 more per dice.

To illustrate this in practice, assume that twenty men are firing at an enemy unit 19 in away from them. This is not the maximum range, nevertheless it is over medium range so it will be necessary to deduct 3 from each dice. The four dice are thrown together and they fall 6: 2: 3: 4. Deducting 3 from each dice—the first volley (6) results in three enemy casualties; the second volley (2) results in no enemy casualties; the third volley (3) similarly results in no enemy casualties and the fourth volley (4) results in one enemy casualty, giving a total of four enemy killed. Had the enemy been sheltered behind a wall, then only the first volley would have caused any casualties, when two men would have been killed, because 1 more has to be deducted from the dice to allow for the shelter of the stone wall.

When firing on cavalry, it is considered that some bullets will hit the horse whilst others will hit the rider. It is not unreasonable to claim that horses can frequently carry on after being hit by a bullet that would lay low a man but, for our purposes, there is a more practical aspect to the matter. We do not have very many cavalry in our game and those which are present are very precious so that to lose them with only one round of firing from a regiment is not only frustrating but unrealistic. Hence half-casualties allow the cavalry to take some fire but still remain a potent force.

The guns of these times were muzzle-loaders and unrifled, a far cry from the precise and accurate pieces of mechanism in use today. Therefore our rules have to make allowances for these imperfections, and this is done by throwing a dice each time the gun is fired. When the range is between 18 in and 36 in it is necessary to throw a 6 on the dice to score a hit; from 9 in to 18 in a 5 or 6 and below 9 in a 4, 5 or 6. Having secured a hit, then throw another dice—its total represents the number of men killed by the shell. Again, count half-casualties for cavalry and also for men sheltering behind a stone wall or in a building. When two forces contact each other so that hand-to-hand fighting, or a mêlée, results, it is decided in the following manner. If it is a straightforward affair between two bodies of infantry, then one dice is thrown for each group of five men and the number of enemy casualties is HALF the total score of the dice thrown. Thus, in the case of twenty Red infantrymen fighting fifteen Blue infantrymen—the Red side throws four dice and perhaps scores 4: 2: 1: 5, which equals 12 and represents six enemy dead. The Blue force throws three dice and comes up with 6: 4: 5, making a total of 15 or seven enemy casualties (when totals are uneven, always level them down). Mêlées between two forces of cavalry are fought in exactly the same manner, but a different situation prevails when the conflict is between a force of cavalry and a force of infantry. If the cavalry have charged into the foot soldiers, they have achieved a certain 'shock' effect by the sheer weight of their horses and the force of their charge. This is represented by adding 1 to each of the dice thrown for the cavalry. In addition, in a mixed mêlée cavalry are counted at two points each whilst infantry count as one point. Thus, in a mêlée between ten cavalry and ten infantry, the cavalry would total twenty points and would throw one dice for every five points, or four dice in all. The unfortunate infantry only counting one point each, making a total of 10, will only throw two dice (one for every five points). When the dust has died down, it could be seen that the cavalry had thrown four dice showing the following scores—2: 6: 6: 3, making a total of 17 or eight enemy infantrymen killed. The two dice of the foot soldiers fell at 5 and 3 making a total of 8, or four points-worth of cavalry casualties, which equals two men. This leaves the cavalry in triumphant possession of the field with eight men remaining against the two infantrymen who, if they have any sense, will run away as quickly as they can. But under our rules they have a very remote chance of showing incredible valour and the highest standard of morale!

Without some way of ending a mêlée, the conflict could drag on unrealistically for two or three moves, so it is not unreasonable to assume that, at the end of a bout of hand-to-hand fighting, one of the contesting forces will become disheartened through losses and turn to flee. In making the decision as to which side suffers this fate, we are really testing the standard of morale of the two opposing forces—perhaps the most potent of all factors in warfare, as recognised by Napoleon who is credited with saying that, 'In war all is mental, and the mind and opinion make up more than half the reality.'

It is not necessary here to go too deeply into this factor of morale in relation to campaigns with model soldiers but, in the author's experience, the best wargames are those in which troops are forced to manoeuvre and conform for reasons of good or bad morale rather than by being unrealistically decimated by fire-power.

To return to that pair of trembling infantry confronted by the looming figures of eight huge cavalrymen. Whether or not they stand or run is decided by each general throwing a dice and multiplying the number of men he has remaining after a mêlée by the score on the dice. Thus if the general commanding the cavalry throws a 1 for his eight survivors, then their morale figure is 8; if the two infantrymen throw a 5 or a 6 then they have triumphed over incredible odds and, by sheer force of example (aided by a lot of luck), they have seen the cavalry off! However, it is more realistic to assume that the cavalry would throw an average score of 3, giving them a total of 24 whilst the infantry's similarly average score of 3 gives them a total of 6. Then common sense prevails, and the two infantrymen make for the hills.

3
Stepped-up Situations

IT MIGHT be tempting for the novice to jump in at the deep end and endeavour to mass thousands of troops on a huge terrain and embark upon a slightly scaled-down Battle of Waterloo. The more sensible beginner discards any such idea, being aware that he is totally unequipped to handle such a vast array at the beginning of his wargaming career. He can find consolation in knowing that one of the joys of campaigning with model soldiers is that really interesting battles can be fought with very small numbers of soldiers and need last only a short time. Those huge battles that run through Saturday and Sunday sometimes tend to drag and have patches of boredom during their lengthy career. They are also immensely time-consuming in organisation, fighting and clearing away afterwards, although such a large-scale battle can be attractive when fought as a 'special' event perhaps two or three times a year.

The wargamer should aim at building up his armies and his experience at one and the same time in progressive steps, starting with a small-scale affair and culminating in a really big campaign involving hundreds of infantry, cavalry and guns. There is, in fact, virtually no limit to what could be achieved and even a battle involving thousands of troops on the largest terrain conceivable under any roof would be practicable. Such a battle could be laid out and left *in situ* to be fought over a period of days, weeks or even months, if so desired.

However, let us begin at the beginning by setting up some essentially basic scenery on a table measuring perhaps 8 ft × 5 ft. Take six trees and a length of wall, set them up as in the diagram (Fig 1) and mark in the road. Now buy two boxes of

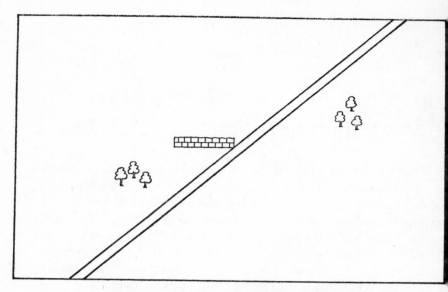

Fig 1   A simple terrain using basic scenery. It is scaled so that 12 in on
the table-top equals ¾ in on the diagram

20-mm plastic model soldiers of the American Civil War period
—one box of Federal infantry and one box of Confederate infantry. Each package will contain about forty-eight little figures
in a variety of action poses, some firing, some running, some
battling hand-to-hand, together with officers and trumpeters.

The American Civil War has been chosen because its fascination is such that, at some stage or other, it is tackled by most
wargamers, probably because it was the first of the 'modern'
wars involving breech-loading guns, rifled muskets, railway
trains, and telegraphs. It also anticipated many tactical and
defensive measures of much later wars. The period, too, is well
covered by readily available books and literature which tell the
campaigner about uniforms, tactics and the men who fought in
such famous battles as Bull Run and Gettysburg. Model soldiers
of the Federal and Confederate armies are not difficult to obtain,
and most manufacturers have a range that caters for the wargamer.

Another attraction is that the uniforms are simplicity itself—
the Federal side invariably wore light blue trousers, dark blue
tunics and a dark blue, flat-topped peak cap (kepi). The Confederate forces wore a motley variety of uniforms, beginning the
war in various combinations of grey, supplemented at frequent
intervals by light blue pants taken from dead or wounded
Federal opponents and later becoming ragged but no less efficient soldiers in slouch hats of all descriptions and shapes and
nondescript uniforms in a colour known as 'butter nut'. (This
is almost any shade of brown that may take your fancy because
the homespun material was also home-dyed in vegetable dyes of
varying quality.)

In the early stages of the war both sides put into the field
many glamorously garbed and titled regiments, some of which
wore the colourful baggy red trousers, the short jackets and
turbans of the French Zouaves who had so recently distin-
guished themselves in the Crimean War. The Louisiana Tigers,
men of French extraction from the New Orleans area of the deep
south, seeking to dress like their beloved Zouaves but lacking
the material for the traditional baggy red trousers, manufactured equally baggy pants out of blue and white striped mattress
ticking. The Confederates also were graced by J. E. B. Stuart's
Virginia Cavalry, who were reputed to be similar, both in attitude and temperament, to the Cavaliers who had fought with
Prince Rupert in the English Civil Wars some three hundred
years before. They supplemented their grey uniforms with yellow-
plumed slouch hats, high black thigh boots and gay coloured
sashes—their fighting ability matched their dashing appearance!

The Airfix figures are manufactured in plastic which resembles
the basic colour of their uniforms, thus the Federal figures are
available in dark blue plastic and the Confederates in grey
plastic. The wargamer will need to decide how much or how
little he wishes to embellish these figures by his own painting
efforts—the man who is easily satisfied might well be content
with dabs of flesh-coloured paint on faces and hands. So let us
assume that we are going little further than this and merely
giving our figures equipment of black or brown, flesh-coloured

hands and faces and, in the case of the rebels, allowing them all to be clad in captured light-blue trousers.

Generally speaking, people who fight campaigns with model soldiers have tidy minds and like their soldiers to be well organised in attractive-looking regiments, finding little pleasure in seeing a regiment scaled down to perhaps twenty other ranks and two officers in perhaps ten different positions. They prefer their twenty men to all be firing, or all to be running, or all to be kneeling, and so to give the desired uniformity of appearance. This is also an aid to recognising a specific unit on the battlefield and makes for easier control during the battle.

Taking as an example the American Civil War figures, let us assume that one wishes to gather together two armies—one Federal and one Confederate—with which to do battle. It is possible to form about ten or eleven different regiments of Federal and about the same number of Confederate infantry from the variety of positions given in the Airfix boxes. This means that the first regiment will consist of all those men in the same firing positions taken from perhaps ten boxes, all those men in a running position will form the second regiment, and so on. Officers gathered from these ten assorted boxes are sufficient ably to control these regiments and when all have been sorted out, two very fine and menacing-looking armies will be facing each other across the table-top.

Wargamers usually amass their forces into regiments of about twenty men with two officers, this being a conveniently scaled-down force which enables one to have perhaps ten regiments of twenty men each with officers on the table for one battle. Such a force, with two cavalry regiments and three or four guns, can give a very fine engagement that will last perhaps three or four hours. For a shorter game use less troops, but if one is having a real gala day and spending an entire Saturday or Sunday on a battle, then by all means throw in twenty or thirty regiments a side. One can then advance, retire, and try outflanking movements with gay abandon, and only after many hours will you survey the tattered survivors and realise that even twenty or thirty regiments are not enough!

In boyhood, most of us collected Meccano, starting with the smallest set and progressing to the larger and more elaborate ones. Each successive set made possible still bigger and better models and when all the sets had been acquired a veritable mechanical Utopia stretched out before the junior engineer. In just this manner, model soldiers, guns and equipment can systematically be acquired to complement those already possessed and build up realistic, miniature representations of the army or armies of any desired period.

No one is going to be content for long with fighting a small battle using under fifty men a side, the majority of whom are in differing positions. The next step, then, is to purchase two more boxes of figures, this time the US Cavalry. Each box contains eleven mounted figures cast in dark-blue plastic so that it will be necessary, in the case of the Confederate forces, for them to be repainted grey or butter-nut. Using the same scenery, set up your infantry regiments but now you may add a troop of cavalry to each side, and with these faster moving 'shock troops', who can create havoc by coming in unexpectedly on the flank, the resulting battle will take on a vastly different complexion.

Another interesting addition can be made to this basic set-up by purchasing a box of American Civil War artillery containing two guns and enough Federal and Confederate gunners to man them. As the box only contains one limber, we will leave it out at this stage but the combination of two or more boxes will enable gun-teams with limbers, outriders and officers easily to be assembled.

Thus for a very small financial outlay, the novice wargamer has assembled enough figures to give him a pleasant little battle on a small scenic terrain. Nor is this the end of it because there still remains an Airfix box of cowboys and the 'waggon-train' set. A few simple conversions, using a razor blade, some glue, small pins, Plasticine, etc, will enable cowboys to be converted into Federal and Confederate troops and take their place on the battlefield. The covered waggons of the 'waggon-train' set are equally easily transformed into military vehicles and the riders in that set can take their place with the other converted soldiers. All very easy and involving no more than a weekend's work to assemble, paint and generally organise these small armies.

And now to begin battling with the two boxes of Confederate and Federal infantry, positioned on a simple battlefield consisting of a road running diagonally across a field from one corner to the other, with a wall in the exact centre of the arena and two small clumps of trees so positioned that each favours one of the two sides. Forty-four warriors will take part in this gladiatorial-type contest to demonstrate the simple rules outlined in the previous chapter. Later, the wargamer can embody some of the factors listed under the Civil War section of this book, so enabling the tactics and fighting methods of the period to be reflected realistically on the table-top.

4
Three Basic Battles

GAME NUMBER ONE

THIS REPRESENTS wargaming at about the simplest basic level, both so far as rules and the numbers involved are concerned. It is a logical foundation upon which to step-up situations in which larger numbers of men and a greater variety of weapons are manœuvred to more complex rules, so affording greater scope and realism.

In this game, moving is carried out alternately—each general throws a dice and the highest scorer has the choice of whether he moves first or allows his opponent to do so. The general who moves second has the right, when within range, to fire his infantry weapons first. To simulate the re-loading of their rifles, infantry are not permitted to fire in the same move as that in which they actually move.

Both generals laid out their troops on the base-line of the battlefield and then each threw a dice to decide the order of moving. The Federal general had the highest score and decided to move first. Both generals moved their two regiments directly forward for the full 12-in distance permitted them.

Move 2. The Confederate general threw the highest dice and decided to make the first move. He sent his right-hand regiment diagonally forward to the shelter of the clump of trees, while his left-hand regiment moved directly forward to within 9 in of the stone wall. The right-hand Federal regiment moved directly forward in the direction of the stone wall while the left-hand regiment moved diagonally to their left towards the grove of trees in which the Confederates were sheltered.

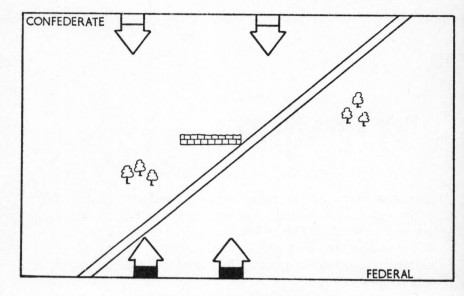

FIG 2   GAME 1   Battlefield layout

Move 3. Having won the dice throw, the Federal general moved first, sending his right-hand regiment directly up to the shelter of the stone wall, while his left-hand regiment moved forward to within about 14 in of the clump of trees. The Confederate general decided not to move either of his regiments, thus giving them the right to fire. All four regiments were extended in single line, and only the two left-hand groups of five men of the Confederate right-flank regiment were in a position to fire on the advancing Federal left-flank group. They threw two dice which came up 4 and 1. At that range, it was necessary to deduct 3 from each dice which resulted in one Federal infantryman being killed. The Confederate left-flank regiment fired upon the Federals who were behind the shelter of the stone wall at a range of about 5 in. They threw four dice for their four volleys, resulting in scores of 6: 6: 3: 5. From each of these dice it was necessary to deduct 1 for range, plus 1 for the protection that the Federals had from the stone wall. When the smoke had died away the Federals were seen to have lost twelve men—a sad blow!

Move 4. Having thrown the highest dice, the Federal general moved first. He moved half of his left-hand regiment to their right behind the shelter of the stone wall, leaving the other half of the regiment to fire, together with the eight remaining infantrymen already in position behind the wall. Facing them on the other side of the wall stood the left-hand Confederate regiment, still at full strength. The right-hand Confederate regiment moved diagonally forward towards the flank of the Federal infantry.

As both sides had men entitled to fire because they had not moved, each side threw a dice to decide who had the right to fire first. The Federals won and fired four volleys over the wall at their Confederate enemies—they scored 2: 2: 3: 5. With 1 deducted per dice, this resulted in eight Confederate infantrymen killed. The survivors fired back three volleys at the Federals, score 3: 3: 3. Deducting 1 for range and 1 to allow for the protection of the wall, this meant that three Federal infantrymen had been killed.

Move 5. Winning the right to move first, the Confederate right-hand regiment charged diagonally into the flank of the left-hand Federal regiment sheltering behind the wall. The other Confederate regiment stood their ground whilst their opponents, the Federal infantrymen of the right-hand regiment, remained in position behind the wall. Unaffected by the mêlée on their left, fifteen Federal infantrymen fired three volleys across the wall, scoring 3: 3: 3: which, with 1 deducted per dice, equalled six Confederates killed. The remaining eight Confederate infantrymen fired two volleys, scoring 1: 1; which caused no losses to their opponents.

Now came the first mêlée of the battle as the Confederate infantry surged triumphantly into the unsuspecting flank of the left-hand Federal regiment. To represent the benefit gained from a flank attack, the Confederates were allowed to add 1 to each of the dice they would throw in the mêlée to represent the result of hand-to-hand fighting. Thus, the Confederate's four dice (one dice for every five men involved) resulted in a score of 6: 4: 3: 1; to which was added 1 more per dice giving a total of 18. Half of this figure represented the total number of Federal dead. In reply, the ten Federal infantrymen involved in the mêlée threw two dice scoring 4: 2; equals 6, or three men killed. Thus the Federals, having lost more men in the mêlée, had to decide the state of their morale, to see whether they stood and fought on or fled. As they only had one man left they were in a very unenviable position and in spite of throwing a 6 on the dice this, multiplied by their one man, gave them a morale total of 6 whereas the nineteen men (seventeen infantrymen plus an officer and a standard-bearer) also threw a 6 giving them a morale total of 114. This meant that the Federal infantryman turned and ran back 12 in towards his own base line.

Move 6. On this occasion, the Federal general was successful in throwing the highest dice and decided to move first. In a daring attempt to save the battle and, at the same time to save themselves from a further attack into their flank by the right-hand Confederate regiment, the remaining Federal infantrymen scrambled over the wall and attacked the line of the Confederate infantrymen facing them. In his turn, the Confederate general moved his right-hand regiment, triumphant from their mêlée victory of the last move, around to the Confederate side of the wall to join the fierce mêlée about to take place. Neither side had any stationary men entitled to fire.

With twenty-eight men, the Confederate force considerably outnumbered the nineteen Federal infantrymen. Nevertheless the latter put up a great fight with their four dice coming up 6: 5: 3: 2, equalling 16 and representing a loss of eight men to the enemy. The Confederate general had five and a half dice to throw which resulted in 6: 3: 2: 2: 2: 2: or a total of 17 which earned him eight Federal casualties. This meant that the Confederate force had twenty men remaining while the Federal force was only eleven strong. Dice were thrown by both generals to determine the state of the morale of their troops. The Confederate general threw a 5 which, multiplied by the twenty men he had remaining, gave him a morale figure of 100, but the Federal

Page 53 (above) 'Troop Blocks' set out in position on a wargames terrain, before being replaced by the actual troop formations they represent; (below) a scenic board, with spare symbols and space for noting casualties etc

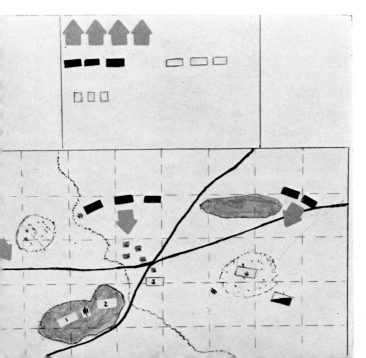

Page 54 A Roman Legion emerges from its fort to engage an opposing horde of Gauls

general, unlucky to the last, threw only a 2 which, multiplied by the eleven men he had remaining, gave him a morale total of only 22. Hence the Federal general had to retreat and, there being no chance of saving the battle, it was deemed a great Confederate victory.

GAME NUMBER TWO

For this battle each side was reinforced by a squadron of cavalry, formed of an officer, a standard-bearer and nine troopers. Thus each force consisted of two regiments of infantry and one squadron of cavalry, a total of forty-four infantry and eleven cavalry.

Instead of the alternate moving of the previous game, this battle was conducted on a simultaneous-moving system in which, after the initial laying out of troops (Fig 3), each general moved his troops at the same time, both beginning from either the right or left flank of the table. A toss of the dice determined from which side the moving should take place at the start of each move.

The generals having laid out their forces as shown on the plan, surveyed their enemy and then prepared for their first move. A dice score of 1, 2 or 3 meant that the moving took place from the general's right; a dice score of 4, 5 or 6 indicated that both commanders had to move from their left. The first move was from the right and at once it could be seen that the Confederate commander had reacted violently to the threat of the left-flank Federal cavalry squadron, for the Confederate cavalry wheeled sharp right towards the right flank of their force. The Federal commander sent his cavalry directly forward, followed on their right by the two infantry regiments, each in two ranks.

Move 2 took place from the left and, even at such an early stage, the pattern of the game became apparent. The Confederate commander had already permitted his Federal opponent to dictate the tactics, in that the Confederate troops were conforming to the Federal movements rather than initiating plans of their

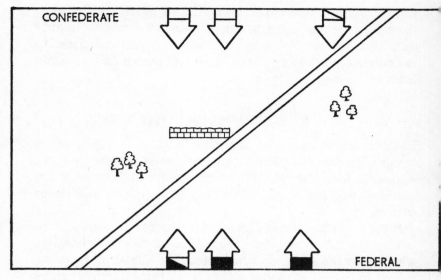

CONFEDERATE

FEDERAL

FIG 3   GAME 2   Battlefield layout

own. The Federal cavalry again advanced and swung slightly so
as to face in towards the Confederate right flank at about the
level of the stone wall. The Confederate right-flank infantry
regiment turned half right to face them with their cavalry
sheltering behind and slightly to their left. The left-flank Confederate infantry regiment moved forward and conformed to
the flanks of its companion unit. The two Federal infantry regiments moved directly forward. Thus, already the Federal commander had managed to pin down two-thirds of the Confederate
force by the threat of his cavalry squadron, whilst his two infantry regiments were moving forward to outnumber the sole
left-flank Confederate infantry unit. Here we see a cardinal
principle of wargaming—to pin down a greater strength than
your own at one point whilst out-numbering the enemy in
another area of the battlefield.

In Move 3 all movement had to begin from the left, which
added to the troubles of the Confederate commander who had
to look anxiously over his shoulder as he moved his left-hand

regiment because of a threatening left-flank Federal cavalry threat. Now thoroughly subservient to the tactics of the Federal commander, the Confederate general turned the left half of his left-flank regiment to face the outflanking threat coming from the right-hand Federal regiment which had moved diagonally to the right of the wall. At the same time the Confederate commander allowed the inner half of his left-flank regiment (in single line) to remain facing forward over the wall against the massed fire of the left-hand Federal regiment. In order to meet the outflanking threat of the Federal cavalry which had gone directly forward and were now swinging to come in on to his flank and rear, the Confederate commander did not move his right-flank regiment but swung his cavalry up to engage the Federal cavalry.

For the first time in the battle, infantry regiments within range of each other, and who had not moved, were able to fire. Dice were thrown to determine the order in which this firing took place.

The half-Confederate regiment behind the wall threw a 6.

The Federals facing them threw a 2.

The two Confederate volleys (less 1 for being within 6 in and 1 because the Federals were behind the protection of the wall) fell as follows 1: 5—each dice, less 2, equalled three Federal infantry killed. Against them the four volleys of the left-hand Federal infantry unit came up 6: 5: 5: 2 (deducting 2 from each dice) resulting in the disastrous total of ten Confederate infantrymen killed, or the almost complete wiping-out of that part of the regiment.

In another part of the field the two cavalry squadrons were locked in fierce combat. Having equal numbers of eleven men each, both Federals and Confederate cavalry threw two dice each to determine the damage that each did to their opponent. Having moved in to make contact with each other, each side was granted a bonus to represent the shock-effect of horses and riders crashing into each other—this was done by adding 1 to each dice. Thus the Confederate cavalry threw two dice and, confirming the Confederate general's suspicion that it was just not his day, each dice came up a 1! To each dice 1 more was added making a total of 4, or two Federal cavalry killed. The Federal general, who knew that it *was* his day, threw two dice which came up 6 and 5. These, plus 1 per dice for shock-effect, made a total of 13, or six Confederate cavalry killed. It seemed unlikely that the morale of the Confederate cavalry would be very high after such a disastrous mêlée but when the dice were thrown to discover the morale-state of both sides, the Confederate commander came up with a 6 which, multiplied by the five men he had remaining, gave him a morale total of 30. The Federal general, with nine men left, had three in hand so he was quite satisfied to throw a 4 on the dice which, multiplied by his nine men, gave him a morale rating of 36. Thus, the Confederate cavalry were forced to retreat, and being very near their own base line they rode off the field, never to return.

Move 5 was again from the left. To cope with the threat to his flank and rear from the victorious Federal cavalry, the Confederate commander ignored his left-hand flank regiment and turned his right-flank regiment to face the cavalry. The Federal commander did not disappoint him and flung his cavalry directly forward into the line of grey-clad Confederate infantry facing them. Having moved, the Confederate infantry were not able to fire upon the charging horsemen. Still displaying tactics of the highest order, the Federal commander swung his left-hand infantry regiment into the rear of the Confederate regiment, already with their hands full of cavalry. The Federal right-flank infantry regiment did not move and so were able to fire upon the unmoving half-strength Confederate unit facing them. The bad luck which had dogged the Confederate commander ever since he had surrendered the initiative at the beginning of the game continued here—the Federal commander won the dice throw for order of firing and was thus able to pour four volleys into the unfortunate Confederate infantry. The four volley dice came up 6: 5: 5: 1. Each dice deducted 2 for range, giving a total of ten Confederate infantrymen killed. As this completely wiped out the Confederate unit, no return fire greeted the Federal infantry, who had completely cleared their flank off the field.

At the other end of the stone wall the twenty-two-strong Confederate infantry unit faced nine cavalrymen attacking frontally and nineteen infantrymen attacking their rear. As these forces were mixed, it was necessary to represent the superior strength of cavalry fighting against infantry—this was done by counting the cavalry as two points each so that they totalled eighteen points, whilst the infantry counted one point each, the nineteen infantry totalling nineteen points. This gave the Federal forces a total of thirty-seven points and as the Federal infantry regiment had caught the enemy in the rear, it was decided that this great advantage should be simulated by adding 1 to each dice thrown for the infantry. Throwing a total of $7\frac{1}{2}$ dice (one dice for every five points) the Federal attackers scored: 6: 6: 5: 5: 5: 5: 1, plus 1 each added to four of the dice to represent the rear attack of the infantry. Thus the total Federal score was thirty-seven points, half of which counted as enemy dead giving a total of eighteen Confederate infantrymen killed.

The twenty-two Confederate infantry were deemed to be entitled to throw $4\frac{1}{2}$ dice which came up 5: 2: 2: 1, totalling eleven points, half of which (five points) represented Federal dead. This was worked out at one cavalryman (two points) and three infantrymen (three points).

Discovering the state of morale of both sides was a mere formality. The five remaining Confederate infantry threw a 3 which gave them a morale figure of 15 whilst the twenty-four remaining men of the mixed Federal cavalry and infantry force threw a 3, which gave them a morale figure of 72. This meant that the five Confederate infantrymen showed the greatest of good sense by scampering from the field and, as their other regiment and their cavalry had already departed the field or been killed, there was no question but that they had suffered a crushing defeat.

Although fought with very small numbers, this little battle developed into an almost classical example of how and how not to fight a wargame. Throughout, the Federal commander displayed intelligence by using simple and effective tactics after the early realisation that he was dictating the course of events. The

Confederate commander allowed himself to be overawed from
the very first move and committed the great wargaming sin of
allowing a smaller force to pin down a larger one. At no time in
the battle, after once surrendering the initiative, was the Confederate commander ever really in with a chance. This little
affair, if studied closely and analysed, can point the way to becoming a successful table-top general with a great reputation for
notable victories.

GAME NUMBER THREE

The same forces that opposed each other in the previous battle
were again in opposition, each side reinforced by a horse-
drawn limber and gun with a crew of four gunners. This meant
that the strength of each of the two armies was two infantry
regiments, each composed of an officer, a standard-bearer and
twenty privates; one squadron of cavalry formed of an officer, a
standard-bearer and nine troopers with one gun and crew.

Besides stepping-up the game in numbers, the method of
moving the troops was improved. Instead of the simultaneous
move with possible disaster resting on the luck of the dice in
deciding the side of the field from which the move began, a
simultaneous move system was used which involved the prior
writing-down of movement instructions. This is the method currently in use among practically all experienced wargamers, certainly in Great Britain. It has the merit of providing a certain
degree of concealment of intention and movement, at the same
time adding realism in that no general can suddenly move a formation to face an unexpected threat. Its disadvantage is that it
nullifies the on-the-spot initiative of both officers and men who
might react rapidly to a sudden onslaught from an unexpected
direction. With table-top experience, however, comes the ability
to build into a set of rules the factors necessary to give a certain
degree of latitude of movement when orders written prior to a
move do not take visible threats into account.

To illustrate this simple although complex-sounding method,
a replica of the actual movement orders written by each oppo-

nent in this battle will be given during the account of the move in question.

With the now familiar battlefield laying bare before them, both the Federal and Confederate generals thoughtfully sucked their pencils as they wrote down the initial layout of their troops. Finally, each had written the following:

Federals

1st Infantry unit	Opp. trees
2nd Infantry unit	Left flank—opp. wall
Cavalry	On right flank
Gun	Opp. right end of wall

Confederates

1st Infantry unit	On left wing
2nd Infantry unit	On right of above unit
Cavalry	Far right flank
Gun	Right of 2nd Infantry unit

Transposed on to the battlefield, this resulted in the initial dispositions as shown on the plan (Fig 4).

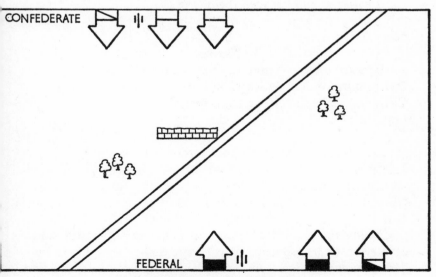

FIG 4   GAME 3   Troop dispositions

Move 1

Federals

1st Infantry unit	Move 12 in forward
2nd Infantry unit	do
Cavalry	Move 18 in forward
Gun	do

Confederates

1st Infantry unit	Move 12 in forward
2nd Infantry unit	do
Cavalry	Move 12 in forward
Gun	do

Reflected on the battlefield, these written orders resulted in the Confederate force advancing with two infantry regiments alongside each other on the left flank, each in single line. On their right was the gun and on the extreme right flank the cavalry squadron. Facing them, the Federal cavalry advanced on their right flank and the two infantry regiments moved straight forward with the gun between them (Fig 5).

Move 2

Federals

1st Infantry unit	Move 12 in forward—face ½ left
2nd Infantry unit	Move 12 in forward
Cavalry	Move 18 in forward
Gun	do

Confederates

1st Infantry unit	—Move 12 in ↗
2nd Infantry unit	— do
Cavalry	Move 18 in forward
Gun	Move left 15 in (Fig 6)

Already the game had begun to open up with both sides' cavalry on their respective right flanks, each attempting to outflank the enemy line. The Federal infantry were in single line across the near centre of the battlefield, the gun slightly in ad-

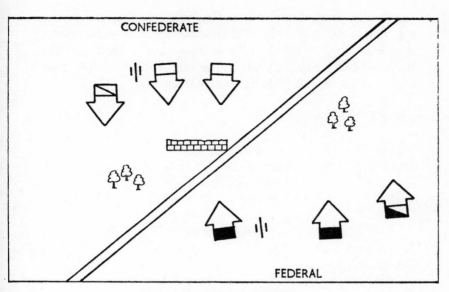

FIG 5   GAME 3   MOVE 1

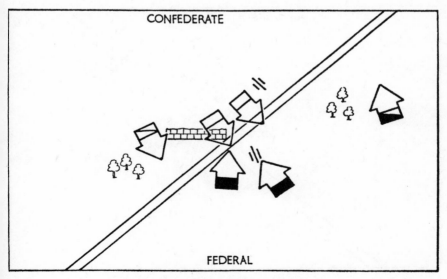

FIG 6   GAME 3   MOVE 2

vance of the two infantry regiments, the left flank unit approaching the stone wall. The Confederate infantry were moving diagonally with their right flank forward and their left flank back (refused) with their gun at the end of their left flank, possibly to counter the Federal cavalry threat.

Move 3

Federals

1st Infantry unit	Move 12 in ↖
2nd Infantry unit	Face cavalry
Cavalry	Move 18 in↖
Gun	Fire

Confederates

1st Infantry unit	Stay still
2nd Infantry unit	do ↓
Cavalry	Move ↓ 18 in
Gun	Unlimber—fire on cavalry (Fig 7)

Both guns remained still; the Confederate infantry did not

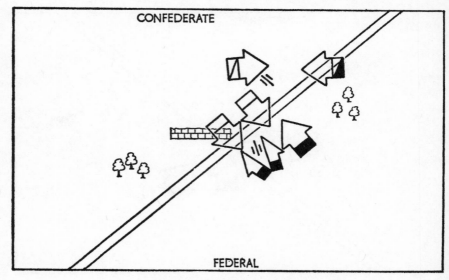

FIG 7   GAME 3   MOVE 3

move; the Federal infantry moved forward with half the left-hand unit facing its left flank. Now came a development which, at first glance, seems to indicate that the Confederate commander had learned nothing from the previous battle. The Federal cavalry had swung out wide to their right, covered by the Confederate gun (guns were allowed to be laid on target at the conclusion of the move, after all movement had taken place). It seemed as though the Confederate commander had again allowed the initiative to be taken by his opponent and that he was conforming to the Federal cavalry movements by repeating his move of the previous game when he swung his cavalry from one flank to another to meet a threat from opposing horsemen. However, the battle was in its early stages and it remained to be seen whether this was a tactical error or a wise move.

The Federal commander won the throw that decided order of firing for artillery and his gun fired at 12-in range on the left-hand Confederate infantry regiment. Requiring a score of 5 or 6 to obtain a hit, he threw a 3 and subsequently missed. The Confederate commander fired on the approaching Federal cavalry at 14-in range and scored a hit by throwing a 5. He then threw another dice to determine casualties and scored a 4, which indicated that four Federal cavalrymen had been killed.

Not having moved, both Confederate infantry regiments were able to fire. The right-hand regiment firing at maximum range (deducting 3) also had to deduct another 1 because their Federal opponents facing them had the protection of the wall. The four Confederate volleys resulted in scores of 1: 4: 4: 3—with 4 deducted from each dice this resulted in no Federal casualties. The left-hand Confederate regiment fired two volleys on the Federal gun at 13-in range and scored 2 and 5. Deducting 3 from each dice for range, this meant that two of the Federal gun crew had been killed. The remaining two Confederate volleys were fired at the infantry regiment facing them, also at long range so that 3 had to be deducted, plus 1 more to allow for the protection of the wall. Dice fell 4 and 4, thus causing no casualties to the Federal infantry.

Move 4

Federals

1st Infantry unit	Stay still
2nd Infantry unit	do
Cavalry	Charge towards gun
Gun	Stay still

Confederates

1st Infantry unit	Stay still
2nd Infantry unit	do
Cavalry	←charge
Gun	Re-lay and fire (Fig 8)

Neither side moved either guns or infantry so that they were all available to fire upon each other. Dice were thrown for order of firing and, having thrown the highest score, the Federal gun led off at 12-in range. Needing a 5 or 6 to secure a hit, a 5 was thrown and the next dice tossed for casualties came up 2 so that two of the left-hand Confederate infantry regiment were killed.

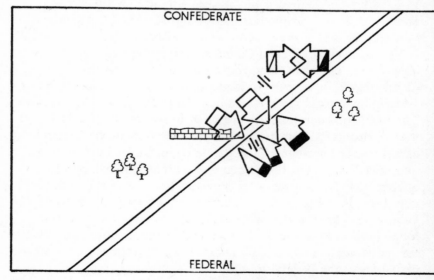

FIG 8   GAME 3   MOVE 4

The Confederate gun fired on the right-hand regiment of Federal infantry and missed. In return, the same regiment fired four volleys, scoring 4: 6: 2—with 3 deducted from each dice for range this gave a total of five men killed in the left flank of the Federal infantry regiment. They, in their turn, fired back their three remaining volleys and proved they were not daunted by their losses by scoring 6: 6: 6. Deducting 3 from each dice for range, this resulted in nine men killed from the right flank of the Confederate infantry regiment. Next, the right-flank Confederate infantry unit fired and scored 1: 4: 1: 5—deducting 3 for range and 1 to allow for the protection of the wall, this resulted in one infantryman being killed in the left-hand Federal regiment. Finally, that same regiment fired its four volleys and scored 4: 5: 6: 6—deducting 3 from each dice for range, this scored nine hits on the right-hand Confederate regiment. As a result of this round of firing, all four regiments were beginning to look somewhat diminished.

Notwithstanding the fact that his cavalry squadron was down to seven men, the Federal commander (no doubt recalling his great cavalry triumphs of the previous battle) decided to charge home on the opposing full-strength Confederate cavalry regiment. Encouraged by their superior numbers, the Confederate cavalry also charged, so that a cavalry mêlée took place with both squadrons adding 1 to each dice to represent the shock-effect of the two forces charging into each other. With eleven men, the Confederates were entitled to throw two dice—lamentably they turned up 1 and 2 which, with the 2 added for shock, gave a total of five points' worth of Federal cavalry dead. In their turn, the seven-strong Federal cavalry squadron were entitled to throw 1½ dice which resulted in 2: 2: plus 1 for shock-effect, giving them a total of five points, the same score as their opponents. It was decided that both sides should lose two men each, which resulted in the Federals having five men left and the Confederates nine. Now came the crunch as both sides threw the all-important morale dice. Throwing a 3, the Confederate general looked smug, conscious that his opponent would need a 6 to better him. But it was not the Federal general's day and he threw a 1, giving him a morale rating of 5, which was hopelessly outclassed by the 27 morale rating of the Confederates. In con-consequence, the Federal cavalry were forced to retreat to their own lines for the full distance of their move.

Move 5

Federals

1st Infantry unit	Back to right of gun
2nd Infantry unit	Stay still
Cavalry	Retreat a full move
Guns	Stay still

Confederates

1st Infantry unit	Stay still
2nd Infantry unit	do
Cavalry	Charge infantry
Guns	Move 6 in and fire (Fig 9)

Seeing his beaten cavalry retreating towards him (although the cause of the retreat occurred in the previous move, the re-

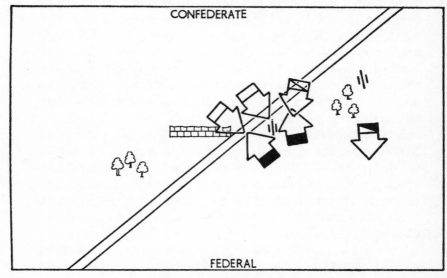

Fɪɢ 9   GAME 3   MOVE 5

treat itself consumed the present move) the Federal commander realised that his right-hand infantry unit had their flank dangerously exposed to the free-to-move Confederate cavalry. Therefore he swung them back, pivoting on the left end of the regiment. At the time this may have seemed the right thing to do, but in the event, it was a disastrous step, because in doing so they forfeited the right to fire. When the Confederate cavalry moved forward as ordered to engage this regiment, their move-distance of 18 in was insufficient to allow them to make contact, so they remained poised threateningly within 6 in of the infantry.

Now the commander moved his gun and the remainder of the infantry stayed firm so that firing was able to take place. The subsequent dice throwing resulted in the following order of firing:

(1) Confederate right-hand infantry regiment,

(2) Federal left-hand infantry regiment,

(3) Federal gun,

(4) Confederate left-hand infantry regiment,

(5) Confederate gun.

(1) Confederate right-hand infantry regiment with three volleys remaining scored 1: 2: 2—deducting 3 for range and 1 for the protection of the wall meant that they had killed no one.

(2) Federal left-hand infantry regiment (four volleys) scored 6: 6: 4: 2—minus 3 for range gave them the score of seven men killed from the right-hand Confederate unit facing them.

(3) The Federal gun, firing at close range on the Confederate cavalry, registered a hit and then scored two cavalrymen killed.

(4) The Confederate left-hand infantry regiment (down to two volleys) concentrated their fire on the crew of the Federal gun. They scored 6: 1—minus 3 for range equalled three hits on the Federal gun crew which killed them all off, leaving the gun useless without any gunners.

(5) The Confederate gun fired on the retreating Federal cavalry with the greatest possible success, registering a hit with a throw of 6 and then scoring six casualties and completely wiping out the Federal cavalry unit.

Move 6

Federals

1st Infantry unit	Stay still
2nd Infantry unit	do
Cavalry	All dead
Gun	No crew

Confederates

1st Infantry unit	Move 12 in→
2nd Infantry unit	Move 12 in→
Cavalry	Charge infantry
Gun	15 in (Fig 10)

The Confederate commander ordered his gun to be limbered up, it then swung round left to appear in the right rear of the Federal forces, where it un-limbered and prepared to fire in the next move.

Both Confederate infantry regiments moved forward only

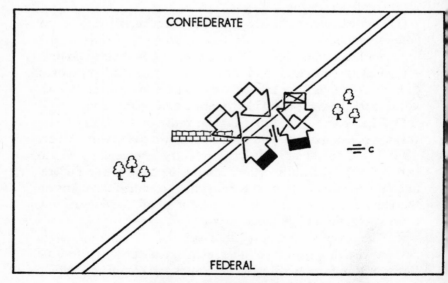

FIG 10   GAME 3   MOVE 6

Page 71 Thirteen hundred years later, the English archers of Henry V form up in characteristic formation in front of a ruined castle to defeat the flower of French chivalry

Page 72 The 37th Foot, the Hampshire Regiment, march bravely forward to assault the redoubt defended by a gun and infantry of the French Royal Roussilon Regiment

their right-hand regiment, coming under fire from the Federal left-hand regiment of infantry who scored 6: 6: 3: 1—minus 2 for range caused nine Confederate infantrymen to be killed. The Federal right-hand regiment was, as expected, charged by the Confederate cavalry but the Federal commander had reason to face this with some confidence as he had two volleys to fire at close range (minus 1 only). His confidence was misplaced because his infantrymen, no doubt upset by the sight of the enemy cavalry, made the rock-bottom score of 1 and 1, so that they did not kill a single cavalryman!

And so, jubilant at their freedom from loss, the seven Confederate cavalrymen crashed into the eight Federal infantrymen, the cavalry adding 1 per dice for shock-effect. The seven cavalry at two points each totalled fourteen points against the eight points of the Federal infantry, so that the Confederates were entitled to three dice. The dice score reflected the elan with which they had charged, coming up 6: 6: 5 which, plus the 3 added for shock-effect, gave them a total of 20, half of which were Federal casualties so that the entire Federal infantry regiment was wiped out. In return, the Federal infantry scored 5 and 3, which gave them a total of eight points, half of which represented two Confederate cavalrymen at two points each. As only one side remained after this mêlée, there was no need to decide any morale rating.

Move 7

Federals

1st Infantry unit	Half face cavalry—remainder stay
2nd Infantry unit	Wiped out
Cavalry	—
Gun	—

Confederates

1st Infantry unit	Stay still
2nd Infantry unit	⌐ do
Cavalry	12 in
Gun	Stay still (Fig 11)

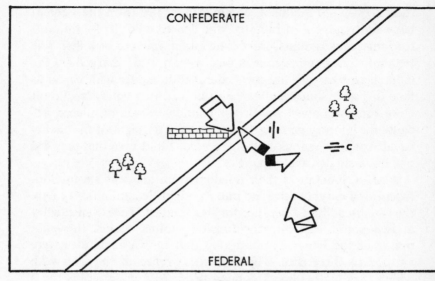

FIG 11   GAME 3   MOVE 7

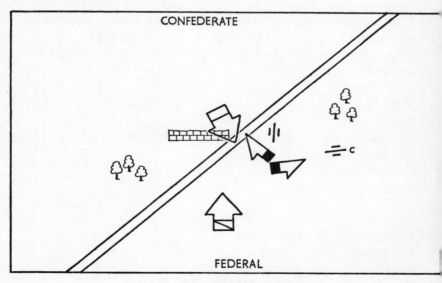

FIG 12   GAME 3   MOVE 8

With only one infantry regiment remaining, his flanks and rear threatened by Confederate cavalry and artillery, to say nothing of a Confederate infantry regiment to his front, the Federal commander had little reason for optimism. However, he made a fight of it by swinging the right-hand half of his remaining regiment to face the threat from the flank, while the men who stood firm fired two volleys at the stationary Confederate infantry to their front, scoring 6 and 3—minus 2 for range equalled five Confederate infantry killed. In return, the Confederate infantry with only one volley to fire, scored 2—minus 2 for range and protection of the wall gave them a nil return for their trouble. The Confederate gun failed to take advantage of the situation and missed.

Move 8

Federals

1st Infantry unit   —
2nd Infantry unit   Stay still
Cavalry   —
Gun   —

Confederates

1st Infantry unit   Stay still
2nd Infantry unit   —
Cavalry   15 in
Gun   Stay still (Fig 12)

The Confederate cavalry swept round directly to the rear of the Federal position, there to remain poised for an attack in the following move.

The Federal infantry was so formed as to be firing on two fronts and it was therefore considered necessary for them to throw two dice to decide upon order of firing. In the event, that part of the Federal regiment firing at the Confederate infantry won the right to fire first and their two volleys scored 6 and 2—minus 3 for range and 1 for the protection of the wall, which gave a total of four hits on the Confederate infantry. In return, the Confederates fired a single volley minus 2, which scored two

hits on the Federal infantry. That part of the Federal regiment
facing to their right fired two volleys on the gun crew, scoring
1 and 4—minus 3 for range, which gave them one hit on the
crew of the Confederate gun.

Move 9

Federals

1st Infantry unit   —
2nd Infantry unit   Two ranks face cavalry
Cavalry   —
Gun   —

Confederates

1st Infantry unit   Move 12 in left
2nd Infantry unit   —
Cavalry   Charge infantry
Gun   6 in move to right (Fig 13)

To face the threat of the Confederate cavalry, the Federal

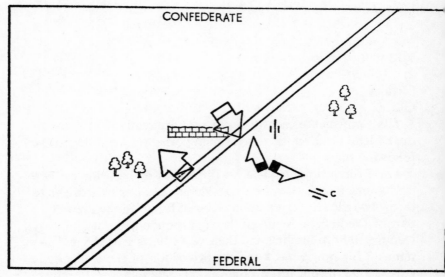

FIG 13   GAME 3   MOVE 9

commander swung inwards half of the men facing the wall to join with half of the regiment facing the gun who had similarly swung back, thus forming a 'V' with the point facing towards the cavalry. This meant that he had one volley to fire forward at the Confederate infantry. It scored 2—minus 2 for range and protection of the wall, and so caused no casualties to the Confederates. The other volley fired on the gun crew and scored 3—minus 2 for range, and resulted in one more of the gun crew being killed. Last in order of firing, the Confederate gun, no doubt shaken by the nagging fire that was diminishing the crew, missed completely.

The Confederate cavalry charged forward into the Federal infantry who, because they had moved, did not have the opportunity of firing upon them. In the resulting mêlée the five cavalry (totalling ten points) scored 5 and 1, plus 2 points for shock-effect, giving them a total of eight points or four Federal infantry killed. The ten Federal infantry engaged in the mêlée scored 6 and 2, giving them a total of 8 also. This resulted in two Confederate cavalry being killed. In the resulting morale throws, the Confederates with three men left threw a 3, which gave them a morale rating of 9, while the six remaining infantrymen threw a 4, which gave them a morale rating of 24. Thus the Confederate cavalry were forced to retreat.

Federals

1st Infantry unit	—
2nd Infantry unit	Half face infantry—remainder stay
Cavalry	—
Gun	—

Confederates

1st Infantry unit	Stay still
2nd Infantry unit	—
Cavalry	Retreat
Gun	Stay/fire (Fig 14)

In the last move the Confederate infantry moved to their left, and reduced to one volley only, they stayed put and fired on the

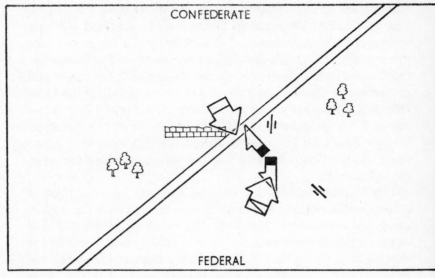

Fig 14   GAME 3   MOVE 10

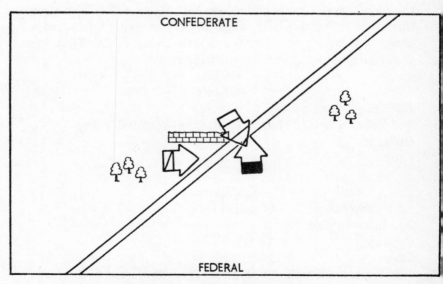

Fig 15   GAME 3   MOVE 11

Federal infantry facing them. The Federal infantry, still facing two ways at once, turned half of their force towards the infantry while the remainder stayed facing the gun.

Order of firing gave the gun the right to fire first—it registered a hit but only succeeded in killing one man. Next the Confederate infantry fired their volley, scoring 6 and, as they were on the wall so depriving the Federal infantry of its protection, only 1 was deducted, which gave them five Federal infantrymen killed. The remaining Federal infantrymen fired one volley at the gun, scoring 6 minus 2 for range, which resulted in four of the gun crew being killed, but as there were only two of them left anyway the Confederate gun was unmanned.

Move 11

Federals

1st Infantry unit	—
2nd Infantry unit	Stay still
Cavalry	—
Gun	—

Confederates

1st Infantry unit	Stay still
2nd Infantry unit	—
Cavalry	18 in
Gun	— (Fig 15)

The only movement that took place here occurred when the never-say-die Confederate cavalry swung round to come in towards the left flank of the Federal infantry.

Order of firing resulted in the Confederates firing first, their single volley scoring 4 minus 1 for range, giving them a total of three Federal infantry killed. That part of the Federal force facing them threw a single volley, scoring 4 minus 1 for range and 1 for the protection of the wall afforded to the Confederates, giving them a total of two Confederate infantry killed.

Move 12

Federals

1st Infantry unit   —
2nd Infantry unit   Charge cavalry
Cavalry   —
Gun   —

Confederates

1st Infantry unit   Charge
2nd Infantry unit   —
Cavalry   Charge
Gun   —   (Fig 16)

At this stage, both generals were imbued with a 'do-or-die' spirit which caused every survivor recklessly to attack his neighbour. The Federal infantry charged the cavalry at the same time as the cavalry charged them! The Confederate infantry on the far side of the wall, reduced to the commander-in-chief plus one

Fɪɢ 16   GAME 3   MOVE 12

officer, came over the wall and valiantly threw themselves upon
the Federal infantry who were also engaged with the cavalry.
No one bothered about such subtle refinements as firing!

In the rather confused mêlée that followed, the Confederates
totalled six points for three cavalry, three points for a mounted
commander, and one point for an infantry officer, giving them a
total of ten points. The Federals had five infantrymen counting
five points, plus one mounted commander at three points, giving
them a total of eight points. As the larger proportion of the
Confederate force was mounted, it was considered fair and
reasonable to give them a bonus of one point for the shock-
effect of mounted men. Thus the Confederates threw two dice
which came up 6 and 1, plus 1 for shock value, counting eight
points or four Federal infantry killed. In their turn, the Federal
infantry threw one and a half dice and scored 5 and 3, which
gave them a total of eight points and resulted in two Confederate cavalrymen killed. When the smoke and dust had died
away, it was possible to see that the field was now occupied by
the commander-in-chief of the Confederate force, backed up by
one cavalryman and an infantry officer, to oppose the Federal
force of a mounted commander-in-chief plus one infantry
officer.

Deciding that only the best was worthy of his steel, each commander-in-chief made straight for the other and a brief mêlée
ensued in the form of a straight dice throw—highest winning.
The Federal commander threw first and horror mirrored his
countenance as the dice showed a 2! The Confederate commander-in-chief's elation was soon dashed however when his
dice turned up 1, which meant that he was dead and the Federal
commander-in-chief had triumphed.

While this had been going on the remaining troops of both
sides had not been mere spectators. The Confederate cavalrymen and the infantry officer (counting three points) had taken
on the Federal infantry officer (counting one point). It was
decided that the Confederates should throw one and a half dice
to the Federals one. They came up 5 and 2 (for the half dice)
equalling a total of 7 for the Confederates, which meant that the

Federal officer had been killed, whilst he threw a 2 which allowed him to be accompanied to Valhalla by his Confederate counterpart.

All that remained now of the two forces was the mounted Federal commander-in-chief and the Confederate cavalryman. In such a basic style of wargaming, commanders-in-chief possess a Herculean quality which makes them better fighters than any-one else. But in these circumstances it is charitable to assume that perhaps the Federal commander-in-chief was a little older and a little more tired than the belligerent young cavalryman whom he faced because, in the mêlée that followed (a straight dice throw with the Federal commander-in-chief adding 1 for his superior status), the cavalryman threw 5 to the Federal commander's 3 plus 1, so that the superior officer was killed.

And so ended the battle with the Confederate force victorious and the surviving cavalryman no doubt feeling rather like the winner of the men's singles at Wimbledon! Although thoroughly enjoyable, it did not, of course, bear much resemblance to any-thing that takes place on real battlefields, and one does not have to wargame for long before tiring of this oft-repeated occurrence of three men and a gun winning a battle by defeating an officer and a private. The need then arises for more realistic and con-clusive ways of terminating battles and, remembering that Wel-lesley was criticised for incurring 28 per cent losses at the battle of Assaye, we recognise that this might best be achieved by cut-ting down casualties in our table-top battles to convincing pro-portions while still arriving at a definite conclusion. Perhaps the simplest way of achieving this end is to nominate some form of objective. For example, a cross-roads in the middle of the battlefield is required to be commanded by the force that declares itself victorious, or perhaps a village has to be held or a ridge dominated. Rarely in real warfare was the objective of the com-mander the virtual annihilation of the opposing army.

Another realistic but more complex method is to so formulate one's rules that regiments are forced out of battle by diminished morale rather than by excessive casualties. More details of this and of the 'objectives' methods are given in a later chapter dealing with more advanced aspects of battles with model soldiers.

Finally, it is interesting to reflect that had the commanders in the third game been better versed in the hobby, the result of the battle might have swung dramatically to a Federal victory in that very last move! This piquant situation might have occurred because most sets of rules contain a proviso that the death of a commanding general has an extremely detrimental effect upon the force under his command. For example, it might be ruled that if the commanding general falls, all units within 12 in of him must ascertain exactly how much their morale was affected by the sight or knowledge of the death of their trusted commander. For each unit so positioned, a dice is thrown and the units conform to the following ruling:

1. The unit breaks up and flees the field.
2. The unit marches off in good order unless rallied by a general within one move.
3. The unit surrenders.
4. The unit falls back one move and ends up facing the enemy.
5. The unit is unaffected and carries on fighting.

It will be recalled that, half-way through Move 12, the last move of the battle, the Federal and Confederate commanders-in-chief both spurred their horses towards each other and engaged in a combat which resulted in the death of the Confederate commander. It will be remembered, too, that whilst this was going on, the remainder of the two forces were also engaged, so that at the end of the two mêlées all that remained was the Federal commander-in-chief and a single Confederate cavalryman. Had the above rule applied, it would then have been necessary for the Confederate cavalryman to have thrown a dice to decide exactly how he reacted to the death of his commander-in-chief and to being faced by the Goliath-like figure of the enemy commander. Had that cavalryman thrown anything but a 5 or 6, then the Federal general would have been considered the victor and the battle would have gone to him as the sole surviving representative of the force lately under his command.

Of course, this rather reduces wargaming to the status of draughts and the reader must bear in mind that the three games we have fought out were on about the simplest possible level. If the wargamer decides to stay with it and carry on, he will certainly not be content to remain at this level but will steadily progress to more complex methods, so making his games ever more enjoyable, and with the added attraction of still greater realism.

5
How to Construct and Use Scenic Boards

EVERYONE IS familiar with the printed chess problem—that chequer-board diagram showing a few chessmen in a situation to be resolved by the reader of the journal. This representation by numbers, quantities and symbols is known as 'notation' and the purpose of this chapter is to explain how a similar system can be used to illustrate on paper a situation involving model soldiers. The system can be used to demonstrate the initial lay-out of troops before a battle, or to pose a problem for the 'reader' at some later stage in a battle. In either case, the first necessity is an easily recognisable set of symbols and a standard scenic board on which to mark them. Because the written situation will often require to be translated on to the actual table-top battle area, the board must be scaled to the size of that table-top. For example, if the scenic table on which the battle is to be fought measures 8 ft × 5 ft then the most convenient scale is 1 in = 1 ft, so that the board will be 8 in × 5 in, ruled into forty-one-inch squares by thin dotted lines.

The sixty-four square chessboard is simplicity itself, but our forty-square scenic board presents some problems of its own. It has to be a simple diagrammatic map, or plan, showing hills, roads, rivers, villages, and such like, together with other recognisable symbols to denote the type of force—infantry, cavalry or artillery formations—moving over that scenic terrain. A suggested set of symbols for both the physical features and the troops is given in Fig 17.

The first principle to be borne in mind is the positive relationship between the size of the symbols used on the scenic board to

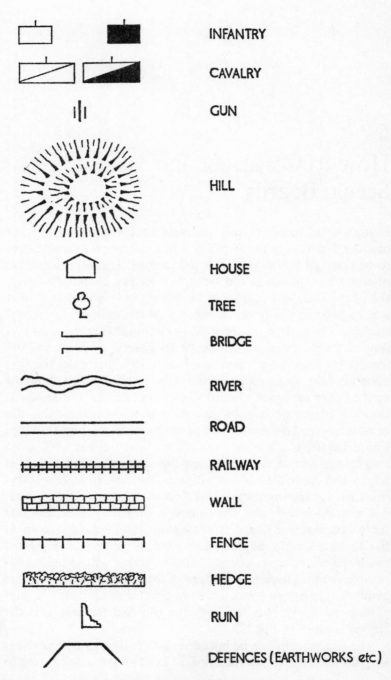

FIG 17   A set of symbols for physical features and troops

their size on the actual scenic battle terrain. For example, a wood that covers one of the inch squares on the scenic board will cover a square foot on the battle-table, whilst a house will have to be fairly small on the scenic board if it is not to become a huge mansion capable of holding a brigade (wargames size) of troops. The scenic board can be used as a miniature diagrammatic plan of the action that is to follow in larger scale—it will be the chart on which one's operational moves are planned initially. The symbols representing the formations of infantry, cavalry and artillery must also be to scale, otherwise the transference of a column of infantry from the board to the table will find them disastrously occupying only perhaps a fraction of the gap defended so well on the scenic board. To avoid such anomalies the symbols for the scenic board must be drawn to the scale of 1 in to 1 ft. The simplest way to explain the manner in which group symbols are used is to set out the procedure of laying the symbols on the scenic board and then transfer the set-up to the larger battle table, making sure that everything is in exactly the same position on the table as it is shown on the scenic board. At this stage, one has the choice of either placing the troops actually on the table, or first representing them by blocks equal in size to the area covered by the troop formation. If it is desired that the battle shall commence at once, then do the former; the second method requires explanation because it is a means of introducing realistic factors of concealment besides considerably speeding up the opening moves of the battle. The troop blocks used on the scenic battle-table are made from pieces of hardboard, balsa wood, cardboard or plastic card in two sets of ten for infantry, four for cavalry and four for guns. Number each one differently and colour one set red and one set blue. The blocks must be exactly the same area as that covered by the formation they represent, and here are some basic dimensions for average troop formations of Airfix 20-mm figures in close order:

approx

Regiment of 20 infantry, plus 2 officers—4 in × 1½ in   (⅓ in × ⅛ in)
Squadron of 13 cavalry, plus 2 officers —7½ in × 2½ in   (⅝ in × ¼ in)
Gun and crew   —2 in × 1½ in   (⅛ in × ⅛ in)
Gun limbered up complete with horses—5½ in × 1¼ in   (½ in × 1/16 in)

The figures in brackets by the side of each set of measurements are the scaled-down size of the symbols to be used on the scenic board, adjusted to a scale of 1 in to 1 ft.

Prepare a chart for each of the red and blue armies, listing numbers 1 to 10 to cover the infantry blocks, 1 to 4 for cavalry and 1 to 4 for guns. Now, against each number on the chart write the name of the unit represented by the numbered block: thus red block No 1 will equal the Ohio Regiment; block No 2 equals the 2nd Virginia Regiment and so on. The cavalry blocks, numbered 1 to 4, could be: Confederate cavalry block No 1 equals the Georgia Cavalry and so on (see Fig 18). The value of this chart is that once the blocks are placed upon the large scenic battle terrain your opponent has little idea of the unit each block represents, other than being able to tell the difference between an infantry and a cavalry block just as, in real life, a person can tell at a distance whether he is looking at cavalry or infantry. By this means it is possible to have 'crack', or élite, regiments positioned as you desire on the battlefield and only your chart tells you whether block No 1 is a veteran regiment, an inexperienced volunteer unit or perhaps militia. Thus it is possible to have regiments of varying quality, morale and fighting powers on the battlefield without your opponent being aware of their precise location. Once the battle commences and the units start moving forward their allocated distance, it is very much easier to pick up a block and move it forward 6 in, 8 in or 12 in as the case may be, than moving perhaps twenty soldiers individually, or even twenty soldiers set in movement trays holding five men each. As soon as the blocks come within combat range of each other (either firing-range or mêlée distance) they are removed from the table and the actual regiments substituted for them.

In that section of Chapter 3 concerned with Game No 3, in which each side had two regiments of infantry, one squadron of cavalry and one gun, the plans showing the events of each move are marked with symbols as here described. By studying them, the reader can readily see how larger maps with more extensive scenery and a greater number of troops can be assembled using the given sets of symbols.

Page 89   Wellington himself watches as the British line prepares to meet (and again defeat) the French column and its accompanying cloud of tirailleurs

Page 90 French infantry with a mitrailleuse form up to beat off the combined cavalry and infantry attack mounted by the Prussians

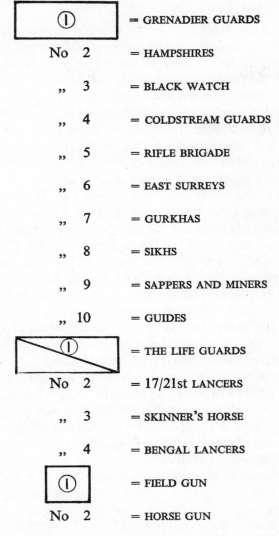

①	= GRENADIER GUARDS
No 2	= HAMPSHIRES
„ 3	= BLACK WATCH
„ 4	= COLDSTREAM GUARDS
„ 5	= RIFLE BRIGADE
„ 6	= EAST SURREYS
„ 7	= GURKHAS
„ 8	= SIKHS
„ 9	= SAPPERS AND MINERS
„ 10	= GUIDES
①	= THE LIFE GUARDS
No 2	= 17/21st LANCERS
„ 3	= SKINNER'S HORSE
„ 4	= BENGAL LANCERS
①	= FIELD GUN
No 2	= HORSE GUN

FIG 18   CHART SHOWING UNITS REPRESENTED BY NUMBERED BLOCKS
As the chart indicates, Infantry Blocks Nos 1 and 4 conceal 'crack' Guard troops, whilst Blocks Nos 7–10 are Native units whose morale varies if their white officers are killed. Much the same situation applies with the Cavalry Blocks—No 1 being Guards, and Nos 3 and 4 Native troops. Gun Block No 1 conceals a Field Gun, whilst Block No 2 represents a much more mobile but lighter (and less effective) horse-gun

Just as the chess problems published in newspapers can be worked out by chess addicts during spare moments, such as when travelling by train, so the enthusiast who fights battles with model soldiers may similarly amuse himself with an easily constructed kit. Armed with this, he will be able to work out by moving, on a board, those symbols that represent troop formations. Symbols representing scenery are not moved.

The first requirement is a sheet of Plastikard, and a thickness of ·030 in will be found quite satisfactory. This Plastikard can be obtained from Bellona, or most hobby shops stock cellophane-wrapped kits of Warneford's Polypaque. These inexpensive packs contain two sheets each 13½ in × 8¼ in. The only other requirement is a card of 'Sasco' shapes, obtainable from office equipment shops and stationers. Known as Sasco Visual Planning products, these coloured shapes and tapes are specially designed for use as symbols in chart-making. They come in a wide variety of colours, shapes and sizes, such as circles, squares, oblongs, strips, diamonds, triangles, and arrows etc. Coated with adhesive on one side, these shapes press on to a glossy surface and may be peeled off and re-used as required; it is possible to write on them and to colour them if desired.

It has already been decided that our wargames scenic board will be scaled 1 in = 1 ft, so that with a table 8 ft × 5 ft in size, the action-part of your board will need to be 8 in × 5 in. However, it is best to give yourself a margin around the action-part so that calculations, casualties and other relevant details can be marked in with a pencil or ballpoint pen. The action-part of the board should be marked in 1-in squares with thin dotted lines—this can be done with a ballpoint or in Indian ink with a fine mapping pen.

The 'Sasco' shapes are put out in packs of about seventy-five shapes at prices low enough to permit them to be used fairly lavishly. Select suitably sized and coloured shapes (brown for hills and roads, blue for rivers, red for houses etc), and cut out terrain-shapes to your requirements. These are stuck on to the scenic board and remain in place for the duration of the battle. Next, cut two sets of symbols to represent the troop formations (to the sizes given elsewhere in this chapter). Select two widely differing colours, such as red and black or blue and yellow. These troop-symbols can be stuck on the reverse side of the board or along the margins at the top and bottom, to be gently eased up with the finger-nail when required.

There are other methods of constructing these wargames scenic boards, two of which dispense with what might be considered the fiddling business of cutting and sticking the 'Sasco' shapes into position. Make a scenic board in exactly the same manner as already described, using the 1-in squares with dots of Indian ink drawn by a fine mapping pen. Using coloured pencils or the wax (chinograph) pencils used for marking maps, draw in on the surface of the board the desired scenic effects of hills, roads, rivers, villages, etc. Using different coloured pencils for the symbols of each of the two opposing armies, mark in their movements as and when they occur and carry on in the same way as if using the adhesive 'Sasco' shapes.

A third method of construction is to have a base-board of Plastikard enclosed in a tightly fitting envelope or cover of transparent plastic material (such as the talc used in map cases). Before inserting the scenic board into its transparent envelope, mark on it the scenic details in their specific colours. With the transparent cover in place, the movement of troops can be marked on that cover without damaging the surface of the actual scenic board by constant drawing and rubbing out. This constructional method also allows of a map, taken from a book or drawn on paper, being inserted inside the transparent cover and resting on the scenic board. The troop movements are marked on the transparent cover in two different colours. This method has the merit of allowing battles to be fought on actual maps rather than on the diagrammatic affairs constructed from 'Sasco' shapes.

When using the wargames scenic boards, the normal rules governing the larger scale wargames table will apply—but with movement and firing distances scaled down. Dice are thrown in the same manner as in the actual battle with model soldiers and casualties are marked in the margin of the board. This can be done on a sort of cribbage system by drawing an oblong for each unit with enough regularly patterned dots inside the oblong so that one dot represents one man in the unit. When casualties occur a line is drawn through a dot until such time as the unit is wiped out or suffers losses enough for its morale-rating to cause it to retreat or even leave the battlefield.

With a little practice these wargames scenic boards can afford many amusing and interesting battles. Being small and portable, they can be slipped into the pocket and brought out to fill in a spare half hour. They can serve admirably as a means of fighting some of the smaller actions that creep into campaigns and which do not really justify an evening on the wargames table or even setting up the required scenery. They can also be used as campaign maps on which to mark troop movements before actually laying out the model soldiers on the full-size wargames table. With the opposing generals privately manoeuvring on their own scenic board, a considerable measure of concealment is obtained before actual contact is made and troops have to be laid on the table.

6
The Lonely Wargamer

So POPULAR has wargaming become over the past few years that only the wargamer who lives in a very inaccessible part of the world need remain solitary in his newly-found interest. He can seek out others with similar interests—a list of Wargamers Clubs is given in Appendix 4—or he can carry on a propaganda campaign among friends, relatives or workmates until one at least takes the bait. But until that occurs, the new recruit may have to plough a lone furrow while he acquaints himself with military history in general or that specific period in which he is interested, learns of the armies, arms, equipment and tactics of the period, and industriously assembles his armies ready for battle.

At times he may feel rather lonely, and may even become discouraged but this is a hurdle which has to be surmounted and a true appreciation of the pleasures of solo battles with model soldiers will be of the greatest value in aiding the solitary wargamer to win through.

Solo battles with model soldiers have much to recommend them, far more than might appear to be the case when first they are contemplated. They should not be looked upon as the inadequate last resort of the man unable to find an opponent, or as rather pointless manoeuvring of model soldiers across a hastily erected battlefield with the same general commanding both sides and knowing exactly what each is going to do. Instead, the solo wargame should be approached in a thoughtful and methodical manner, recognising that it is a facet of the hobby requiring even deeper thought and interest than regular scheduled battles with an opponent. Unlike chess or the better-known board games, or even football or golf, the rules governing wargaming are peculiar to each individual wargamer. They stem from his interpretation of the manner in which the troops fought in the period in question—no man living has withstood a full-scale cavalry charge so it is highly improbable that anyone who has only read of such a shattering experience should be capable of formulating rules that realistically evaluate its effects and reactions. The man who has a passion for cavalry and likes to throw them around the wargames table with all the elan of a Murat or Kellermann will violently disagree as to their capabilities when charging the square of infantry set up by the wargamer who is devoted to the infantryman and who has a pretty stolid temperament anyway! No rules are foolproof, and opposing generals with differing interpretations of them have frequently been known to hold up the game whilst they argue bitterly. No such clash of opinion will distract the wargamer who fights solo battles. For him there is the reasoned solution to difficult situations, solved with a harmony and timelessness that make the hobby the true relaxation it should be in this tense day and age. Not for him the imperative and sometimes hasty laying out of a battlefield in time for the scheduled 7 pm arrival of his opponent, nor the urgent desire to get on with the battle and secure a result before that opponent has to leave at perhaps 10.30 or some other reasonable hour.

The solo wargamer can take his time mulling over what he intends to do—is he going to reproduce in miniature a reconstruction of some famous battle of the past, does he intend to fight a completely imaginary conflict, or is the engagement to be more in the nature of an exercise designed to test some military theory or tactic? Are the opening shots of some memorable campaign to be fired in this initial battle—because solo wargaming ideally lends itself to the long-drawn-out campaign which is often difficult to fight with an opponent whose attendances may be irregular. Every single soldier taking part in this battle is known to the wargamer, he has painted them all or converted them or even made them from scratch.

In the evenings of preparation that ensue before the battle begins, the solo wargamer can plan the composition of his forces —both armies. Thoroughly he plans the battlefield on which they are to fight, because this is an all-important part of war-gaming against oneself. A badly laid out battlefield can 'channel' troops on to one section of it because a river or an impassable marsh splits the table up into two or three sections. Unless a specific tactical problem is being posed, the wargamer should ensure that the battlefield is reasonably 'balanced' so that a territorial advantage to one army is countered by an equal benefit to the other, and a piece of disadvantageous terrain similarly compensated for. There is great fascination in building up realistic battlefields, with hills and roads and rivers, with red-roofed villages to provide objectives for posed problems, and woods to form colourful obstacles to cavalry or refuges for hard-pressed infantrymen.

Most wargamers get a great deal of pleasure from building up battlefields, and then standing back to survey them with the cal-culating eye for ground possessed by all good commanders. If the terrain has been built 'off the cuff', then the time has arrived to draw a reasonably sized and correctly scaled map on which to mark the dispositions and tactical designs of the opposing armies. Or perhaps the battlefield has been carefully constructed from a map previously drawn, or from a map of an actual battle-field of past days. In either case, the solo wargamer has before him yet another happy evening in which he can mark with coloured chinograph pencils upon the talc or transparent-plastic-covered map.

Solo battles can even be more realistic than normal wargames providing that both sides conform to previously made tactical plans. And here, too, lies the answer to the solo wargamer's most perplexing problem and the charge most often advanced against battles lacking an opponent—their inherent lack of surprise. Because the same commander is manipulating both armies and can counter any tactical ploy that either might make, it is assumed that there can be no element of surprise. This is un-sound reasoning, however, because there are a number of methods by which the solo wargamer can introduce a far greater degree of surprise and concealment than is usual between oppos-ing generals in a normal wargame. Perhaps the simplest system is to make a set of cards, each one the size of a formed-up infantry, cavalry or artillery unit. These should be marked so that there is a predominance of infantry, fewer cavalry and even fewer guns, plus blank cards equal in number to the combined totals of the three arms already mentioned. To use them, let us assume that Red army consists of five infantry units, two cavalry squadrons and two guns. Gather the nine cards representing those formations, together with nine blank cards, and shuffle them. Then, without knowing whether a card is a blank or represents a unit, lay the cards in position on the battlefield in dispositions more or less to your liking. Do the same thing with the opposing Blue army and then carry out the opening moves of the battle until contact is made. When either those cards in contact are turned up and revealed in their true light, or else ALL cards on the table are turned up, a true picture is revealed of the actual battle situation in all its horror (or delight). There is an interesting variation possible with these cards in that one army has the usual quota of marked cards plus an equal number of blank cards, while the enemy are given a tactical advantage by only having half the normal number of blank cards so that their commander has a greater sense of assurance that his troops are actually where he intends them to be. This is a ploy that can be made to counter an army of high morale (ie, that with the most blank cards) facing an army of lower morale.

Another method is to give each unit involved in the battle a specific 'morale-rating' so that an élite or guard unit has a morale value of say three points, while an ordinary line regiment has a morale rating of two points, and a militia or irregular regiment is only worth one point. Using cards marked accordingly, they are shuffled and then set out on the battlefield (without using blank cards) so that neither their own commander nor the enemy initially knows where he is strong or where he is weak. Many variations suggest themselves: the solo wargamer can enlist the services of a friend or other wargamer living some distance from him who will (perhaps by mail) allocate the dispositions of one side, leaving the solo wargamer to handle the other. The cards can be laid in battle positions, or they can be laid out on the respective base lines at each side of the table, moving forward when the game begins with the chance of manoeuvring to adjust any obvious weaknesses or to counter any specific threat.

The use of 'chance-cards' will also introduce variety and interest into a table-top battle. These cards can be strictly tactical or, if the solo wargamer has that sort of sense of humour, they can pose problems of a semi-facetious nature, such as by telling their unfortunate drawer that, 'The Imperial Guard discovered the wine cellar in the chateau where they rested overnight. As a result, only half of them are fit to parade and their fighting efficiency is reduced by 50 per cent.' A tactical chance-card, on the other hand, might read, 'As a result of snow melting in the mountains, the river has swollen and is now impassable', or, 'The bridge at point X, weakened by the military traffic passing over it during the past few days, has collapsed and is out of action for a day', or again, 'Only ten rounds of ammunition per man remain.' Such interesting problems as logistics (the supply and reinforcement of armies in the field) can be reflected by chance-cards and, to that end, the cards can be formulated in graded series so that one set deals exclusively with ammunition, another with rations, another with communications, and so on.

Perhaps the best system, and the one most likely to give an interesting and realistic battle for the solo wargamer, is the use of the general tactical card. In real-life warfare, one army usually attacks whilst the other defends—inversely, battles with model soldiers usually take the form of two armies each attempting to attack the other until one is forced on to the defensive. Of course there are exceptions to this rule, as when two opposing forces meet unexpectedly in an 'encounter battle' and fight desperately for territorial advantages until one side assumes superiority and the other falls back upon the defensive. The use of the tactical card can reflect the offensive/defensive character of real-life warfare.

Make two sets of major tactical cards—one for the attack and one for the defence. The attack cards should bear such instructions as, 'Attack with your main force on your right (left) flank whilst holding the other flank with a smaller force', or, 'Ignore flanks and make a centre thrust', or, 'Make a feint attack with half your army while the other half attempts an outflanking movement (on the map and coming on to the table at a chosen moment).' The defence cards could read, 'Send forward a thin line of skirmishers until the enemy's intentions are known and then make your dispositions accordingly', or, 'Hold strongly the most obvious defensive position on your side of the battlefield while holding a reserve for a counter-attack', and such like. Further sets of minor attack and defence cards should be made out, each bearing a relation to the major card under which they are operating. For example, if the major attack card orders an attack in 'oblique order' (one flank moving forward in strength whilst the other flank is refused or held back) then the minor defence card might say, 'Mass your cavalry behind the refused flank and prevent it being turned.' The major defence card that is coupled with the major attack card of this nature might well read, 'Attempt to hold attacking flank with inferior numbers whilst massing to break through the enemy's centre', and the minor defence card could instruct, 'Ignore refused flank and concentrate in your centre.'

It is essential that all tactical cards are 'keyed' in together so that the anomalous situation does not arise of both forces being ordered to attack or both armies to defend. Instructions can, of course, be of a much vaguer nature than already indicated, so leaving a greater degree of personal responsibility to the general concerned—but remember that this is solo wargaming and you are both generals!

In spite of the pleasures and even advantages of solo battles, the majority of wargamers will no doubt prefer to fight against a live opponent. This usually comes about when two friends battle regularly against each other, or when two erstwhile strangers become friends as they progress within the hobby. Later, perhaps, friends or relatives, seeing the enjoyment to be derived from battles with model soldiers, show interest and wish to take a hand in the game. So the pair grows to three, four or even more, and whilst this can be fun, battles with more than two combatants are apt to produce difficulties. After having been in sole command of his army, a general is loth to share that command

and find to his chagrin that, whilst his flank is moving victoriously forward, the other man's forces are being scattered all over the field at the other end of the table! A group of three wargamers is hard to handle because one has to fight two unless it is taken in turns for two to fight and one to umpire. When two fight one, the single commander has a definite advantage because he knows what is going on all over the field and his troops are more or less doing what he has in mind, whereas a joint command is as good as the weakest of the pair and both generals tend to fight their idea of the battle regardless of what their ally is doing by their side. Four players can make for a pleasant and reasonable game, with each force divided into two and with two men commanding one force against the two men of the other. Even so, this sort of game usually develops into two separate wargames alongside each other on the same table!

When the four turn into five or six then difficulties really begin. Five or six men crowded round a table in a single room form a crowd and, with the best will in the world, they are going to produce:

(a) Noise, even if they only discuss the game, and uproar if they argue.

(b) If two or more smoke, then the odour of tobacco hangs around the house for days (perhaps a minor point but distasteful if the normal occupants of the house are non-smokers).

(c) Upheaval in normal domestic arrangements in that their movements and conversation keep children awake whilst disturbing the routine of a wife or mother who may be barred from her living-room because it is the only one with sufficient space for a wargames table.

(d) Assuming that the host-wargamer extends conventional hospitality towards his guest allies and opponents, refreshments each week for such a group of thirsty and hungry wargamers can mount up to a sizeable drain on the housekeeping money!

Soon it becomes apparent that different arrangements have to be made and inevitably the idea of a wargames club is mooted. And this is by far the best solution to the problem of battles with model soldiers involving four or more wargamers.

The initial aims of the group should be modest and it should always be borne in mind that the prime purpose of the organisation is to fight battles with model soldiers under such arrangements that groups can be accommodated in enjoyable wargaming. After the club has been given a title, it is a good idea to twist the arm of one of its members and persuade him to act as secretary. However loose the official organisation of the group may be (and this is most desirable in the first place) there will be occasions when letters have to be written and inquiries answered —a secretary is required to take the responsibility for doing this. Meetings will have to be arranged and dates fixed—again, this can be done by the secretary, or a fixture secretary can be appointed in addition.

Ideally, the club should try and acquire some premises, even if it be only a large room or cellar. Here a permanent wargames table or tables can be set up, scenery stored and figures left until the next battle. It should even be possible for a battle to be left *in situ* for several weeks if it has not been completed in a single evening. The group can jointly subscribe to magazines of interest and aid, and a circulation list can be compiled so that every member has the opportunity of borrowing them in turn. A small library of books can be accumulated and a club information file can contain details of armies, their uniforms, equipment and tactics. Later, it may be considered desirable to have a small club news-sheet or magazine, though this may possibly come more under the heading of publicity.

This leads to another important aspect of the wargames club —the need to attract new members and so strengthen the group. A club magazine or newsletter can be sent to the local newspaper, which will usually give the club some publicity. It can also be sent to schools, youth clubs and Army cadet groups, etc. A notice of the club's formation, its venue and the times and dates of meetings should be sent to the editors of such magazines as *Wargamer's Newsletter*, *Airfix Magazine*, *Meccano Magazine*, and the like. One enterprising club had some attractive leaflets printed and left them lying around all over town. In another case, an enthusiastic member borrowed, in rapid succession, large numbers of war books from the local library and returned them with a club leaflet tucked in their pages! Incidentally, many libraries have glass cases in which they regularly stage displays of interest to people using the library, and chief librarians are often alive to the attraction of a group of model soldiers set against a background of dust covers from the latest books dealing with history and war. In such a case a small notice about the club should be placed in the showcase, giving the address to which those who are interested may write for further details. Hobby exhibitions and shows are other excellent opportunities for publicising the activities of a club—at the time of writing the Wessex Military Society are planning to refight the English Civil War battle of Cheriton during a hobbies exhibition in Winchester, only a few miles away from the actual battlefield.

But perhaps the strongest reason of all for forming a wargames club is the opportunity it provides for collective projects. Most wargamers laboriously buy, make, and convert, and then paint up an army and its opponents. Sometimes, a more enlightened pair of wargamers will arrange for one to collect the army of one country whilst the other collects its enemy. In a wargames club it is possible for large numbers of troops to be assembled in a very much shorter time so that wargames can get to grips on a large and attractive scale without delay. Taking the Napoleonic Wars as an example, one wargamer can be responsible for the British forces, another for the French, another for the Prussian, another for the Russian, another for the Austrian, and so on. Acknowledging that everyone will want to have a British or a French army, perhaps it can be split up so that one wargamer collects British infantry and French cavalry while another collects British cavalry and French infantry, and so on.

Another type of collective project is for a campaign to be selected, such as the English Civil War, and for each member of the club to be allowed to put into the field two regiments of infantry and a gun, or two squadrons of cavalry and a gun, or one regiment of infantry and one squadron of cavalry and a gun. On the night of the battle (which will be part of a co-ordinated and prearranged campaign) everyone will turn up with their regiments and the Cavaliers (who may have five adherents present), will put into the field their massed forces against the Parliamentarians (who are unfortunately only three on parade). Thus realism is immediately brought to the battle in that it is fought between those men who arrive on the field, as in real life, rather than arranging things so that both sides are reasonably equal—a pernicious custom in the world of battling with model soldiers.

It will not be long before the wargames club of one town will attract the attention of another in a nearby town and a challenge will be thrown out. This is an excellent way to broaden one's interests and learn how the other fellow goes about the hobby, particularly as battles with model soldiers are often apt to become parochial affairs, with each club doggedly sticking to their own sets of rules. When fighting against another club, some compromise must be made—perhaps the home club can supply the rules when entertaining visitors, while the visitors' rules would apply when a return visit is made.

Let us now assume that it is club night and that the first arrivals have set up the table so that two or three battles can take place simultaneously. Three or four members have brought along armies and on one table a modern battle is about to take place, on another a Napoleonic set-to, whilst on the third table a Roman army is about to annihilate some Ancient Britons. As people arrive so they survey the scene and decide the period in which they would like to fight, because this is an excellent opportunity for the man who collects and normally fights with moderns to try his hand at being a Roman commander or the chief of a tribe of Ancient Britons. The evening will go all too quickly and, at its conclusion, the terrain and figures are packed away, the tables folded up and the room left in a reasonably tidy condition.

It sometimes happens that a wargamer discovers a fellow hobbyist ideally suited in every way to be his regular opponent but who lives too far away for them to be able to meet regularly. For such as these, postal battles offer a means by which they can, to some extent at least, pit their wits against each other and so make the best of the situation.

The simplest method of fighting a postal wargame is to handle the strategical side of it by post whilst fighting the actual battles in the respective home towns of each of the two opposing generals. Each general requires an Ordnance Survey map scaled 1 in to the mile. It should be divided into map-reference 1-in squares. One general takes the northern sector of the map, while the other commander is responsible for the southern area.

At this stage is it preferable to work out some sort of narrative giving reasons for the coming conflict. For example, the King of the North is insulted because the Prince of the South has spurned the hand of his daughter in marriage. Or else it can be the old story of one side's lust for territorial gain and the other side's heroic defence of their homeland. Whatever the *causus belli*, each commander should select a specified number of cities plus a capital within his area and give them appropriate names. The capital should be classified as being worth three points and the other cities one point apiece. The object of the campaign can be to capture and destroy five enemy points while retaining five points of your own.

Set the rates of movement at three squares by road or two squares across country per move, a contact being made when two opposing armies enter the same square. The forces thus contacted have the choice of fighting or retiring.

Each commander can have a specified number of armies, say three or four, and each army is of roughly the same strength although not of the same composition. By this it is meant that points values are given to infantry units, cavalry units and guns, so that a commander may be entitled to pick a force of, say, 250 points, with twenty points equalling an infantry unit, thirty points a cavalry unit, and ten points per gun. In this way it is possible to select armies which are suitable for the particular terrain over which they are fighting—in other words, the sensible general will not pick a force strong in cavalry if he is attempting to climb a steep mountain range. The armies can be deployed anywhere within each frontier and the winner of a battle will suffer no loss, while the beaten side is considered to have lost half their army and will have to retreat at least a full square in a rearward direction to their commander's choice. An army so reduced can no longer fight but may retire into a town and stand siege, having to surrender if not relieved within three moves. Or a defeated army could unite with another defeated half-strength army to form one full army. After four battles have been fought, each general may bring on a maximum number of reinforcements equal to one full army, but neither country is allowed to bring on sufficient reinforcements to raise their force above its initial numbers.

The campaign begins when each general makes his initial troop dispositions on his map (ideally covering it with a sheet of transparent plastic and marking in his troop movements by using the symbols given elsewhere in this book—in this way the face of the map suffers no damage and can be used repeatedly). When each general has completed his dispositions, he notes them down on a sheet of paper, giving appropriate map-references to each force, seals the sheet in an envelope and sends it to his opponent. Thus the Red general, having made his own dispositions, will receive the enemy's sealed envelope which he will open and then mark in the enemy's disposition on his map. If a greater degree of concealment is desired, then, instead of sending the enemy the exact troop dispositions, send only a general indication of the area (say one equalling 2 × 2 squares) is given. This would certainly be realistic in that reports would probably filter through from the inhabitants of an area or from scouting patrols that the enemy were known to be in such and such an area rather than in one specific place. Both generals continue making their moves and acquainting each other with them in the manner described until such time as two forces arrive in the same square. Accepting that both sides are prepared to risk battle, the scene is now set for the campaign really to come to life.

In his own home, each general sets up on his wargames table a scenic terrain approximating to that of the map square in which the contact has taken place. He masses his army together and also assembles an army equal in size and composition to that of the enemy. This force is handled either by his normal and

Page 107 A German trench raid against the French—flame-throwers gush amid shell-holes and trenches

Page 108 An assortment of German armour moves forward against lighter British tanks supported by an anti-tank gun and a 25 pdr

regular wargames opponent in his own town, or else the battle is fought as a solo game. And so we have a situation where a contact has been made, and *each* general in his own town has fought that battle. Now it may well be that the Red general in London will secure a convincing victory with his forces, while the Blue general (perhaps in California) has secured an equally convincing victory with *his* forces! So we have an anomalous situation in which both sides have won the same battle. In that case, by some mutual arrangement, the battle will be re-fought in one or other of the towns and the result will then count towards the campaign. On the other hand, if the Red force wins both in London and California, they are now the outright victors and the Blue force has to fall back the specified distance.

And so the campaign continues, with each general making his moves and suffering defeat or glorying in victory, cheerfully accepting the results of battles fought with unknown generals sometimes many thousands of miles away commanding his forces. Reinforcements come in when sufficient losses have been incurred to make their presence legal and it is not very long before each general's map presents a most interesting picture with arrows denoting the movements of his forces, and outflanking movements taking place almost before his very eyes.

There is a great deal of 'branch-line' amusement to be obtained from these postal campaigns. For example, each general can wage a propaganda campaign by sending news-sheets to the enemy telling of the glorious feats and aims of the army of that particular country. Chance cards can be used—a duplicate set being made so that each general has the same number and type of cards from which a specified number must be picked at regular intervals. Naturally, his opponent will not know of the dire (or otherwise) effects of the dictates on these cards and so will carry on his troop movements knowing nothing of what a glorious opportunity he may, perhaps, be missing of winning the war.

All this may appear to be of a rather basic nature, as indeed it is, and the wargamer indulging in postal battles may wish (particularly if his side are losing) to have rather more control over the actual battles. After all, both sides stand or fall in battles in which their cherished forces are under the control of strangers whose tactical ability is unknown to the actual general. So the scheme can be developed to give a greater degree of control to both generals by each sending (when a battle contact has been made) sealed instructions to the wargamer handling his forces. These instructions will be of a tactical nature, perhaps setting out the initial dispositions of the troops (because the general will be fully aware of the nature of the scenic terrain over which the battle is being fought, although he will not know the manner in which that terrain has been reconstructed from the map on to the wargames table). His instructional notes might be brief and say, 'Defend—holding village until opportunity is seen to launch counter-attack', or he might tell his stand-in to, 'Mass behind the ridge and make your attack around its western end.' Admittedly, this method is a little tricky but, if handled with patience and common sense by all concerned, it will certainly liven up the postal campaign and make it well worth the trouble.

Postal battles with model soldiers, then, can be an interesting and novel supplement to a wargamer's regular meetings with his local opponent. It should not be difficult to find someone, at home or even abroad, willing to take part in a postal campaign and editors of most journals concerned with the wargames hobby are usually willing and able to put correspondents into touch with one another.

PART TWO
Which Period do You Prefer?

7
Ancient Warfare

THIS SECOND part of the book briefly surveys warfare, weapons and warriors from the beginnings of recorded history up to the present day. It considers the tactical points of each period, together with the role of different types of soldiers and their weapons, because only with some such knowledge of military history is it possible to select a period for which accurately detailed armies of model soldiers can be assembled and campaigns fought with some degree of realism.

Man could not fight wars, as distinct from individual combats, until he had a weapon to hold and this did not really occur until the discovery of metals in about 2000 BC. Local scuffles or brawls no doubt occurred between primitive tribes or individuals but the absence from the remains of palaeolithic or neolithic man of any tools or instruments designed for fighting alone indicates that war was not an organised or regular activity. The Bronze Age gave mankind the fatal gifts of arms and armour, so making soldiering a trade and fighting a skilled activity.

The first fighting weapon known to mankind was the bronze dagger. Inevitably, it was lengthened to become a sword and attached to a pole to form a spear—all offensive weapons. Man's ingenuity then led him to devise defensive appliances such as the shield, the helmet and the breast-plate. Once started, fighting and warfare took an ever increasing hold upon Man, not only as an instrument of policy or a means of settling problems both big and small, but also as a stimulating and dangerous extension of hunting wild animals.

The earliest beginnings of warfare are shrouded in the mists of antiquity, because it originated before history began to be written. Thus we have no clear picture of how men first came to organise armies or to fight in military formations. Later, carvings and monuments did reveal something of the armaments and equipment of the standing armies of the Empires of the Nile and the Euphrates, but the strategy or tactics of the Egyptian and Assyrian Wars are still a mystery. It is not until the Graeco-Persian Wars of the fifth century BC, that we have any intelligible narratives of battles and campaigns—so this must be the starting point of our consideration of military history.

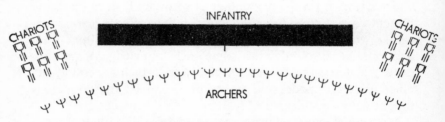

FIG 19   THE EGYPTIAN METHOD OF WARFARE (1500 BC)
First, the archers advanced and fired upon the enemy. Then the spear-armed heavy infantry marched steadily forward while the chariots rolled away at the trot, breaking into full gallop as they approached the enemy

The Persian method of fighting was first to wear down the enemy with arrows and then pick the right moment to launch a devasting cavalry charge against him. The Persian infantry, armed with short swords and small shields, rarely played a prominent part in these battles so that, as time went on, they became weaker than the other two Persian arms—the cavalry and the archers. Nevertheless, these Persian tactics succeeded against every enemy they encountered—until they came to grips with the Greeks, whose pikemen made short work of the Persian infantry whenever they were able to get to close quarters with them.

The typical Greek soldier was known as a hoplite. A heavily armed pikeman, he carried a 6-ft spear and a short sword and wore a helmet, a shield, a cuirass and greaves (leg armour). The backbone of the Greek army, the hoplites were sometimes sup-

ported by small forces of cavalry and archers but it was not until
the arrival of Alexander the Great that these auxiliaries began to
play a major part in Greek warfare. The hoplites fought in a
formation eight files deep, forming a dense hedge of spears
known as a phalanx. This was a formidable formation that was
almost irresistible in a charge. But as time went on, ways were
found of defeating the phalanx. Its weaknesses were a lack of
mobility, so that on rough ground it was easily thrown into disorder and, because it could not quickly change front, its flanks
were vulnerable. Because each man carried his shield on his left
arm and sought instinctively to bring his unprotected right side
under his right-hand comrade's shield, the phalanx always
tended to bear to the right in a charge. This meant that when
two phalanxes charged each other, in each case the right wing
invariably outflanked and defeated the enemy left so that the
battle was decided (or bogged down) by a confused mêlée between the two victorious flanks.

For details of how Epaminondas solved this problem, read
accounts of the Battles of Leuctra (371 BC) and Mantinea (326

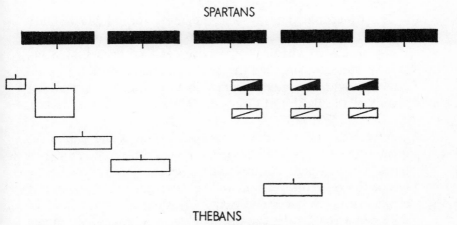

FIG 20   Leuctra 371 BC: The first use of the Oblique Order. Epaminondas
with 6,000 Thebans, defeated 11,000 Spartans by using a tactical method
to be adopted and perfected by Frederick the Great more than 2,000 years
later

BC). Other battles which give a good idea of the method of fighting in this period are Marathon in 490 BC, Thermopylae in 480 BC, and Platoea in 479 BC.

Alexander the Great perfected the almost professional army he inherited from his father, Philip. It contained every variety of fighting soldier of the day, including hired mercenaries and large numbers of highly experienced veterans. An improved phalanx formation was devised in which the pikemen formed up sixteen ranks deep, each man carrying a 12-ft sarissa in place of the 6-ft spear of the Greek hoplite. Then there were the peltasts, light infantry with the same offensive weapons as the hoplite but with much lighter defensive armour. The peltasts had the unique ability to fight in an organised open formation—a skill acquired only through long and intensive training. In addition to archers and javelin men, Alexander possessed an efficient siege train so that he could besiege cities and strongholds. Finally, there were two divisions of Macedonian cavalry formed of the Companions of the King, picked troops who fought on the right or exposed flank of the phalanx, and, in addition, the Thessalian and Greek allied cavalry, lighter horsemen who fought more defensively on the opposite wing.

The Macedonian army was the most formidable military machine the world had known up to that time and with it Philip destroyed Greek independence at Charonea in 338 BC, and his son later smashed the power of mighty Persia. For details of how these highly professional armies fought, read accounts of the Battles of Granicus (334 BC), the Issus (333 BC), and Arbela (331 BC).

Alexander and his armies travelled far afield and one of his greatest battles was with the Indian King Porus in the Punjab, a battle in which elephants took part. There has been considerable discussion over the years as to the rules governing the use of elephants in wargames and only by reading of their success or failure in actual battles can the man who campaigns with model soldiers be able to formulate satisfactory rules to cover their use. Basically, their military value was invariably exaggerated and successful ways of coping with them were soon learned.

Cavalry were useless against them because horses would not face elephants and although the huge animals did great execution amongst densely massed infantry, the pikemen of Alexander's army, although suffering heavy casualties, defeated the elephants by their steadiness and discipline. Nevertheless, the elephants made a great impression on Alexander's generals and they built up an elephant corps which lasted for more than a hundred years. Elephants frequently appeared on the battlefields of the third century BC but gradually fell into disuse as their ineffectiveness became apparent. At the Battle of Zama in 202 BC, the Carthaginian general Hannibal sent about eighty elephants against the Roman lines. Scipio, the Roman commander, drew up his maniples in columns and the elephants, maddened by arrows and other missiles, charged harmlessly down the lanes or were thrown back on the Carthaginian wings, where they threw the cavalry into disorder.

When considering the armies of Rome it must be borne in mind that the period under review stretches from the fourth century BC to the fifth century AD. This entire period, or a great number of eras selected from it, hold the greatest fascination for the man who campaigns with model soldiers and many wargamers specialise solely in the Roman army. In the fourth century BC the Romans, appreciating the advantages of a more open order, abandoned the phalanx formation and entirely recast the structure of the legion. Now it was divided into three sections:

The Hastati, or young soldiers.

The Principes—men of maturer age.

The Triarii—the veterans.

Each section was sub-divided into ten companies known as maniples—those of the first two sections numbering 120 men each and those of the third division sixty men. Attached to these heavy infantry companies were 1,200 lightly armed troops known as velites, and 300 horsemen. The normal strength of the legion was 4,500 men but on occasions it was enlarged to 6,000. For a campaign, four legions were usually raised, each consul commanding two, assisted by six tribunes to each legion, while each maniple was officered by two centurions.

Fig 21   The Roman 'chequered' formation of the fourth century BC

At this period, the legionaries were equipped with helmet, shield, cuirass and short sword. The Triarii carried the hasta, or pike, formerly carried by all ranks, while the Hastati and Principes each carried two javelins (pila). Grouped in a chequered chessboard formation, the maniples in battle could be manoeuvred separately as independent units, filling gaps in the front line and closing up and carrying on the fight if part of the legion was overwhelmed. This ability to continue fighting when penetrated by the enemy derived from the fact that the legion was a flexible phalanx—cohesive yet divisible, fluid and manoeuvreable yet solid and weighty. The Roman method of fighting was first to send forward the light infantry in skirmishing order, then the Hastati charged, hurled their javelins and attacked the enemy with their swords; finally came the Principes and Triarii, following up in support.

Full details of the way in which Roman armies of this period fought are given in accounts of the Punic Wars, when Rome fought a life and death struggle against Carthage. The first of these wars raged from 264 to 241 BC and, whilst the land fighting was indecisive, Rome won her victory on the sea. In the second Punic War, 219 to 202 BC, the Carthaginians were led by Hannibal, acknowledged to be one of the great generals of all time. Battles to read up are the Trebbia (218 BC), Lake Trasimene (217 BC), Cannae (216 BC), Zama (212 BC), Metaurus (207 BC), and the battles which took place in Africa when the Roman Scipio defeated Hasdubal, another Carthaginian general.

It is worth recording that the Battle of Cannae is considered to represent 'the high watermark of military achievement in the Ancient world'. It was 'the perfect battle', in which a strong force was surrounded and destroyed by a weaker one. It has been written that '. . . no later battle has ever quite displayed the same combination of originality in conception and artistry in execution'.

One of the most intriguing facets of campaigning with model soldiers is the opportunity it gives to reproduce and test the effect of one military formation or weapon when opposed to another. Such table-top tests of strength can be used to determine hypothetical situations as, for example, when pitting the English archer and his longbow against Wellington's Peninsular veterans with the 'Brown Bess' musket. One such situation presents itself from the Punic Wars which is well worthy of a table-top trial—that is the Roman legion versus the phalanx. There are three recorded instances of the phalanx being very roughly handled by the legion although it must be admitted that, at this time, the phalanx was no longer at its best. At Cynoscephalae in 197 BC, the phalanx was lured on to rough ground so that it came into action in a disjointed manner with the right wing first advancing and defeating its opponent while the left wing was caught on the march and charged before the men could lower their spears. A Roman centurion marched twenty maniples across the battlefield to take the victorious wing of the phalanx in the rear and decide the battle. At Magnesia in 189 BC, Roman cavalry massed on the right wing routed the enemy left and hit the phalanx in the flank. At Pydna in 168 BC, the phalanx advanced too hastily so that a gap opened in its centre and the Romans charged in.

These faults of the phalanx were due to the length of the sarissa (the pike) being increased to 21 ft, resulting in even greater difficulty than usual in manoeuvring the phalanx because the men had to grasp their huge spears with both hands and were bunched tightly together. Unable to change front, the phalanx went to pieces if assailed in the flank or rear because, whereas Alexander used his cavalry to protect his flanks, his successors relied almost entirely on unaided infantry. On level ground against an adversary who waited to be attacked, the phalanx was still formidable but when confronted with the highly mobile and flexible legion, the phalanx was '. . . like a bull surrounded by nimble toreadors'.

Marius, a Roman general who lived from 157 to 86 BC, was responsible for many military reforms, the most important of them probably being the substitution of the cohort for the maniple, a change prompted by the need to oppose a denser formation than the loose lines of maniples against savage warriors who staked everything on a wild rush and tremendous physical strength. The cohort became a permanent unit of the Roman army, consisting of 600 men divided into six centuries and with ten cohorts to each legion. In battle, the larger cohorts

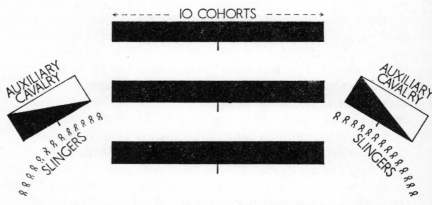

FIG 22   The Roman Legion's Method of Attack: A legion consisted of ten cohorts each of 600 men, divided into ten 60-man groups set out three-deep on a frontage of 2,000 ft. In the attack, the front line of cohorts advanced until, at 200 ft, the first two rows of each cohort ran forward and flung their pila when within 50 ft of the enemy. Next, the rear rank did the same, then the front ranks were among the enemy thrusting with their swords. Meanwhile, the second line of cohorts was preparing to move forward in like manner, while the third line of cohorts waited in reserve, either to be used for 'mopping-up' after victory, or covering the retreat if their comrades were defeated. Throughout, the slingers maintained a hail of stones on the enemy's rear and the auxiliary cavalry drove in on the enemy flanks to bunch and disorder them for the frontal attack

had broader fronts and narrower intervals and were better suited for tactical movements than the maniple.

Among other innovations, Marius abandoned the old distinction between Hastati, Principes and Triarii so that all legionaries were armed alike with the pilum and the short Spanish sword introduced by Scipio. The soldiers were drilled thoroughly in the use of their weapons in the fashion of the gladiatorial schools and the light troops and cavalry were withdrawn from the legion. Rome now recruited her auxiliary forces from the provinces so that Gaulish horsemen, Cretan archers and Balearic slingers were regular units of the Roman army.

When the era of Roman conquest had passed by the second century AD, the legions were relying increasingly on defensive tactics; hand-to-hand fighting was rare, the sword had been discarded for the spear and missile weapons were the predominant arm. In place of offensive strategy, the Romans fell back more and more on to fortification; elaborate systems of earthworks, blockhouses, forts and walls (like Hadrian's Wall in Britain) held back the barbarians.

Later, in the third century, barbarian tribes were invited to serve as units under Roman standards, an innovation which soon destroyed the discipline and efficiency of the legions. In time they adopted the arms, tactics and even the war cries of their ancient enemies and forsook their traditional mode of fighting. By the fifth century, the legion as a military formation was as obsolete as the phalanx.

The barbarians recruited into the Roman service were mostly mounted warriors, and this fact, together with the progressive deterioration of the legionary infantry, brought about a predominance of cavalry in the Roman army. In about 300 AD, Diocletian increased the proportion of cavalry from a tenth to a third, and a hundred years later infantry formed less than half of the total numbers of the army. Their role had similarly deteriorated so that they were now used almost exclusively for scouting, skirmishing and garrison duties. And so the supremacy of the horse soldier was established, a supremacy that was to survive unbroken for nearly a thousand years.

Ancient Battles on the Table-top

So far as wargaming is concerned, the period covered by the title 'Ancient' extends nearly 2,000 years. In this lengthy period, weapons changed from stone to bronze and then to iron. First employed to pull a war chariot, the horse was used a a powerful weapon of war when cavalry began to play their role, and became even more important when stirrups were introduced. The elephant and the chariot made their fearsome impact, only to become obsolete as tactics were devised to counter them. In the same manner, new ideas, tactics and formations first raised and then brought down such military factions as infantry, cavalry, elephants and chariots, all of which, in their turn, had spells as masters of the battlefield.

So it can be seen that it is impossible to draft a single set of rules for the entire period, although basic details can serve as a foundation upon which to build up rule-books for specific eras. It will help, too, to read as much as possible of the soldiers, their weapons and tactics, so that realistic simulations of the manner in which they fought can be transferred to the table-top battlefield. It will be best to start with general rules of a fairly simple nature and, with experience, will come the ability to devise additions to represent the differing races, weapons and tactics of this two-thousand-year period of warfare.

Obviously it is not possible here to discuss every facet of this period in relation to its representation on the wargames table, and what now follows is a collection of hints applicable to the principal warrior-nations covered by the title 'Ancient'.

The Greeks

In the period before Alexander, the Greeks fought in disciplined units of hoplites, bearing long spears. Such a formation was successful more often than not when opposed by loosely organised, less disciplined troops who will need, in a wargame, to decide whether or not they withstood such an assault. Thus, at some agreed point either on contact or just previously, a dice score of 5 or 6 will be needed if they are to stand firm and fight the Greeks. If they *do* stand, then the long Greek spears should be allowed to do some damage *before* the actual mêlée takes place.

Providing the Greeks stood firm, cavalry could not charge home on the well-ordered formation. When the cavalry attack, a Greek dice score of 3, 4, 5 or 6 ensures that they stand. At the same time the cavalry, seeing this firm hedgehog of gleaming spear points, might well baulk at charging into them. So, if the the Greeks stand firm, then the cavalry need a dice score of 5 or 6 to charge into them. Otherwise they swerve away and either turn back or carry on past the formation. If both the Greeks stand and the cavalry attack, then a normal mêlée takes place.

The Macedonians
Developing the Greek hoplite spearman, the Macedonians armed their soldiers with an 18-ft spear (sarissa) and formed them up in a tightly-knit phalanx. Attacking loosely organised troops, this formation was virtually unstoppable. Thus, against such troops in a wargame, the phalanx will *always* break through. Normally requiring to be six ranks deep, the phalanx must be *three* times the depth of formed, disciplined troops if it is to break through. The attack must be delivered over level ground and the flanks of the phalanx must be protected by supporting troops—then the defenders need to make certain dice scores to fight back:

    1 or 2—the defending formation is completely shattered, all men flee.

    3 or 4—the defending formation is broken, the phalanx grinds through to the limit of its move. A mêlée is fought with the attackers adding 50 per cent to their dice score.

    5 or 6—the defending formation holds firm and a normal mêlée is fought.

Although cavalry cannot *frontally* charge a phalanx, the flanks of the formation are very vulnerable, particularly when it is engaged in a frontal attack. If a phalanx is attacked when it does NOT have its flanks protected at the same time as it is frontally engaged, then it is necessary to decide by dice-throw whether it survives:

1 or 2—the phalanx breaks up and flees.

3 or 4—the phalanx breaks up but a mêlée is fought with the flank-attackers gaining a 50 per cent bonus to their score.

5 or 6—the phalanx changes front to meet the attack and a normal mêlée is fought.

The Romans

Because they were highly disciplined, the Romans were able to fight in a looser formation than the Greeks. Their tactics varied considerably throughout their long period of military supremacy, but basically the Romans were swordsmen, with each legionary carrying two heavy javelins with long iron heads (pila). Just before contact, the pila were discharged in two volleys, then the disorganised enemy were charged with the sword. This can be represented by giving the pila a range of 3 in on the table, anyone hit requires a dice throw of 5 or 6 to carry on fighting *providing* he has a shield or is wearing armour. Men lacking such protection are killed immediately.

The various 'tribesmen' enemies of Rome, such as the Gauls, the Britons, the Huns and the Goths, relied on crushing the legionaries' formations by sheer weight of numbers, speed and irresistable courage. They charged powerfully, but if the first (or possibly a second) such charge failed to break the formation, then the barbarians were discouraged and defeated. It was a method of warfare almost exactly duplicated some fourteen hundred years later when British defensive formations broke the incredibly brave onrushing Zulus and Dervishes.

Elephants

Originally, war elephants were used by Eastern nations, and Alexander the Great came up against them when he invaded India in 326 BC. The Greeks used them on a number of occasions and they were employed by the Carthaginians against the Romans, and by Hannibal in the Punic Wars.

Elephants were most effective against soldiers meeting them for the first time, particularly if those troops were not particularly well disciplined. Trained, professional troops could handle them—the Romans, as we have seen, opened ranks and allowed the animals to pass through. Frightened by the noise, goaded by arrows and spears, its driver killed or wounded, a stampeding elephant was as great a menace to its own side as to the enemy.

For the wargamer, there are two elephant models of a scale suitable to blend with Airfix figures—the African elephant from the Airfix Zoo set and the 'baby elephant' made by Wm Britains. Both have to be adapted for wargaming by being fitted with a howdah and crewed by a driver (mahout) and archers. The howdah is really only an appropriately-sized 'box', of card or plastic-sheet, fitted on to the elephant's back with the aid of 'saddle' of thick Plasticene. It can be embellished by having shields and arrow-quivers fixed to its sides. The driver and crew can be converted from various suitable Airfix figures.

Elephants rarely 'crashed' through infantry at high speed, they were more inclined to amble forward at infantry-charging speed. They tended to force units out of line rather than trample them, although this occurred once a unit broke and ran. They were also vulnerable to archers, who could pick off the mahout (driver) and the bowmen in the elephant's howdah.

Cavalry rarely attacked elephants because horses did not like going near the big animals; a line of elephants was sometimes used as a barrier against cavalry. Horses cannot discriminate in their dislike of elephants so that cavalry can only work with 'friendly' elephants if they have become accustomed to the presence of these 'ponderous pachyderms'. It requires a dice score of 4, 5 or 6 for cavalry to work with elephants.

Camels

These ungainly animals were used by Eastern nations and have a high 'nuisance' value on the wargames table because not only do they count as 'extra-heavy cavalry' but their smell causes horses to bolt! At the beginning of the battle, decide wind direction and give your camels a 'smell-range' of 6 in down-wind, 3 in cross-wind, and nil up-wind. Any horses coming within this range will bolt unless they score 4, 5 or 6 on the dice (each horse throwing individual dice).

Chariots

The Egyptians and the Greeks used light chariots drawn by two horses with a crew of driver and bowman. Not so extensively used by the Greeks, chariots were the backbone of the Ancient Egyptian army. The Assyrians, Gauls and Ancient Britons used heavy chariots drawn by two or four horses and carrying a driver with two men.

Both the 'Roman' and the 'Ancient Briton' sets of Airfix figures include chariots and crews. Apart from being used for racing in the circus, there is little record of the Romans ever using chariots in actual warfare. Nevertheless, those in the Airfix sets can easily be converted into the type used by other nations of the Ancient World.

Usually divided into squadrons, chariots deployed on the wings of an army or worked in front of the main body of infantry. Chariots became less effective as trained and disciplined heavy infantry appeared on the battlefield. The Greek phalanx, armed with long spears, and the Roman cohort, formed solid blocks capable of resisting the attacks of chariots, or else they were sufficiently well trained to open lanes for chariots to pass through.

Less disciplined troops would find it difficult to repel a mass attack of chariots providing the vehicles could keep moving. But once 'bogged-down' by masses of infantry, the driver and archer in the chariot would soon be overwhelmed.

Value of Armour and Shield

An important feature in both Ancient and Medieval periods was the protective value of armour and shields, and it is essential that this should be represented realistically on the wargames table. Obviously, men wearing armour or carrying a shield are far less vulnerable, both to missile-fire and to hand weapons in close combat. To simulate this protection, a dice is thrown for each casualty after they have been hit or brought down in a mêlée.

A dice score of 6 means that an unprotected man has only been slightly wounded and is able to carry on fighting. A dice score of 5 or 6 enables a man wearing armour OR carrying a shield to fight on. A dice score of 4, 6 or 6 allows a man to carry on if he is wearing armour AND carrying a shield.

8
Medieval Warfare

IN THE thousand years stretching from AD 400 to 1400 it is fair to state that all the best-known armies of Europe and Asia were composed mainly of cavalry, their infantry forces being badly equipped, trained and disciplined. Into this category fall Justinian's forces and those of the later Byzantine Empire, and the ravaging hordes of Goths, Huns, Mongols, Saracens and Turks right down to the feudal levies of medieval Europe.

The reason for this dramatic reversal of military roles could have been the co-incidental deterioration of the legion infantry of the later Roman Empire, but it cannot be accounted for by any really marked superiority of the horseman over the efficient foot soldier. The use of stirrups revolutionised the fighting efficiency of horsemen, while mail-clad and mounted knights could, without danger to themselves, cut down any number of unarmoured peasants. The situation altered dramatically, however, when the infantry were in a good defensive position, or were armoured. The most likely reason for the supremacy of cavalry over infantry during this period was the general inability of infantry to take the offensive, an action of which only highly trained and disciplined soldiers are capable. Between the decline of the Roman legion and the advent of the Swiss pikemen, Europe could field no force of infantry possessing these qualities.

The outstanding feature of medieval warfare was its indecisiveness, inherent in its highly mobile armies and impregnable castles. Battles only took place when both sides were confident of victory; it was impossible to force battle upon a mounted army which did not want to fight as it simply galloped away and shut itself up in its castles. It took a prolonged siege to reduce a medieval castle, far longer than the normal six weeks duration of a campaign, beyond which few medieval armies were prepared to remain together as a fighting force. This meant that the besieging force had melted away long before the garrison of a castle had reached the point of surrender through starvation. In the field, manoeuvring was non-existent and warfare became a matter of marching straight for the enemy and then taking part in a vast and sprawling cavalry mêlée.

To gain some idea of the modes of fighting of the different cavalry armies of this lengthy period, study of the following battles is recommended:

Adrianople (AD 378), a Goth cavalry army overwhelms Roman legionaries.

Tricameron (AD 535), Belisarius leading Byzantines to victory over the Vandals in Africa.

Lechfeld (AD 955), where the German heavy cavalry beat the Magyars mounted bowmen.

Hastings (1066), the victory of Norman feudal cavalry over Saxon armoured infantry.

Manzikert (1071), a mounted Byzantine army under Romanus are defeated by the Seljukian Turks, a fierce race of mounted bowmen from central Asia.

Mohi (1241), the rout of a European feudal army by mounted Mongols under Subutai.

The Crusades (1096 to 1270), battles between two types of cavalry army—the mailed chivalry of Europe and the mounted bowmen of Asia.

The Hundred Years War (1337 to 1453) was mainly made up of three outstanding campaigns: Crecy (1346), Poitiers (1356) and Agincourt (1415). In each of these battles, dismounted English armies, formed largely of longbowmen and armoured men-at-arms, defeated the mounted feudal chivalry of France. This war ended in French victory through guerrilla tactics and the use of the cannon, which robbed the longbow of its supremacy. It ushered in a new era in military history when firepower was transferred to a progressively improving weapon of almost limitless capabilities.

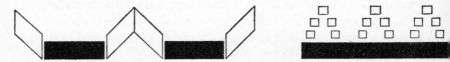

FIG 23   English longbowmen of the Hundred Years War period formed up in a 'harrow' or wedge-shaped formation allowing the maximum number of bows to be brought into action in a converging fire. This formation was gloriously victorious at Crecy, Poitiers and Agincourt. Smaller forces may have formed-up their archers in front of the men-at arms in five groups each of five men

This era of military history cannot, however, be left without reference to the Swiss pikemen who were as much responsible as any other factor for terminating the cavalry age before gunpowder had had time to make its devasting mark upon warfare. Not since the days of the Roman legion had there been any infantry who could do more than stand on the defensive until the Swiss pikemen developed a form of infantry warfare which could take the offensive even against cavalry. Armed with 21-ft long pikes, and in formations resembling the Macedonian phalanx, the Swiss infantry advanced in broad deep columns which did not lose their alignment even when the men who formed them wheeled or changed direction. Only the most exhaustive drilling and highly developed discipline, plus the introduction of marching in step to military music or the sound of the drum, made possible these complicated manoeuvres. In battle, the Swiss army formed up in three divisions: the advance guard, the main battle, and the rearguard. One division attacked and held the enemy while the others assailed him in flank and rear; when thrown on the defensive the Swiss formed large squares with crossbowmen and halberdiers at the corners. They had no hesitation in charging mounted troops, and in this manoeuvre they were almost invariably successful

Soon the Swiss became the most formidable fighting men in Europe and every important army enrolled them as mercenaries. They won famous victories at Morgarten in 1315, at Laupen in 1339, at Sempach in 1386, at Näfels in 1338, at Granson in

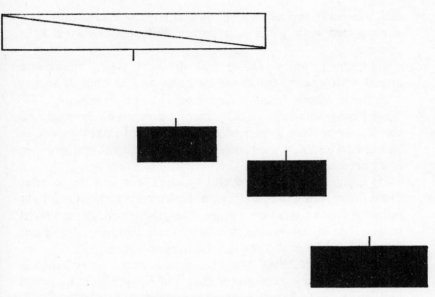

FIG 24 The attack of the Swiss pikemen: Each division marched steadily forward, pikes thrust out in front. The first division made straight for the enemy line, the second division were ready to move across to counter any outflanking movement, whilst the third division, sometimes halted, waited to see the effect of the first attack before going on to throw themselves in at the point most likely to rout the enemy

1476, at Morat and Nancy in 1477. Inevitably, new military developments and armaments made it impossible for pikemen to triumph unaided and at Marignano in 1515 the days of the Swiss pikemen came to an end when their charges were broken by French cannon.

Medieval Battles with Model Soldiers
The armies of the time were probably formed of the following types of soldiers:

(1) Armed Peasants. Without armour of any kind, armed with scythes, bills, spears and clubs, some possessing poor-quality bows.

(2) City Militia. Raised by most Continental countries of any

size, especially in Italy and Flanders. Almost exclusively, they were armed with pikes and spears and possibly wore light armour.

(3) Light Infantry. Of several different types, they were armed with spears, crossbows or bows and in English armies with the longbow. Usually, they wore little or no armour.

(4) Heavy Infantry. Usually men-at-arms, with armour and shields. They were armed with spears, swords, axes, maces or halberds. The Swiss used only the pike or halberd and wore no body armour.

(5) Light Cavalry. Of several types. There was the hobilar (used for scouting and raiding on the Marches) who was a light infantryman mounted on a pony. Then there were the mounted bowmen or crossbowmen, the latter fired dismounted and not from the saddle as did the Asiatic horse-bowman.

(6) Medium Cavalry. Mounted men-at-arms, lesser knights, squires, etc. They wore chain mail and carried shields, sword and lance. Their horses were unarmoured.

(7) Knights and Barons. When mounted, they were classed as heavy cavalry. They wore plate armour and shield, and were armed with lance, sword and axe. Their horses often were partly armoured.

This was a period when organisation and discipline were at a low ebb, so that fighting was between individuals rather than regiments or groups. Usually, a feudal army consisted of bands of men grouped around their feudal seigneur, the groups being loosely banded into an army under a king or prince. If a leader, be he king or seigneur, fell, then his men might lose heart and give way. To reflect this factor when battling with model soldiers it is suggested that a dice should be thrown on the fall of a king or leader (or if his standard is captured).

- (a) When the fallen man is a feudal lord, test the fighting state of his group.
- (b) On the fall of a king or leader ALL groups and lords need to be tested to see whether they stand or run.

A simple form of test would be to throw a dice for each lord or group as the case may be:

1. Throw down arms and surrender (the lords and knights are taken prisoner, lesser ranks are always slaughtered).
2. Throw down arms and flee.
3. Fight to hold present positions but refuse to advance.
4. Make a fighting withdrawal.

5 or 6. Fight on.

This was a period when the armourer's art reached its peak and everyone, except the lowly peasant or bowman, wore chainmail or plate armour giving varying degrees of protection. The method of representing the protective effects of armour described in the section dealing with Ancient warfare can equally well be utilised in this period of warfare.

Among the many specific types of warriors covered by this lengthy period, the horse-bowman is one which lends itself ideally to realistic use on the wargames table. These mounted archers, usually of Eastern origin, were extremely adept at firing their short composite bows whilst riding at full gallop. Excellent examples can be discovered in the descriptions of the battles between the Crusaders and the Saracens. To represent such features on the wargames table, horse-archers are given the faculty of split-moving. That is to say they move a part of their move, fire their bows, then complete their move OR fire their bows and then move. Thus a realistic representation would be to divide their move distance up into three—for example, an 18-in move is divided into three sections of 6 in and the horse-bowman is allowed to fire at the end of the first 6 in on a target which he happens to be passing, then at the end of 12 in on another nearby target, and finally at the end of 18 in on the target nearest which he finishes his run. On the other hand, he can move, say, 9 in out of his 18 in—fire—and then complete the remaining 9 in to ride back into safety. Although the miniature figure representing the horse-archer will obviously have the man pointing in a certain direction, rules should be made to allow them to have an all-round field of fire. That is to say that the imagination of the wargamer must be stretched to allow the bowman to swing in his saddle and fire to right or left, or even behind him if he is retreating.

There was a vast disparity between the power of the weakest and the strongest soldiers during this period. The peasants or militia, armed with their poor, makeshift weapons, stood little chance against the heavily armoured men-at-arms and practically none at all against the armoured horsemen. Rules should, therefore, be formulated to give the weaker men a very slight chance of upsetting the odds against them but, basically speaking, cavalry would always charge home against them, with the sole exception of the English longbowman. When charging into infantry, the heavily armoured horseman did not go in at a fierce gallop but trotted in and pushed his way through by sheer strength, so that rules must be devised to represent this crushing or pushing aside of foot soldiers who, according to their fighting spirit, would attempt to retaliate by hooking the rider off his horse or hamstringing the animal.

The Swiss pikemen and their redoubtable formations have been mentioned. When battling with model soldiers, these can be represented by adapting the suggestions covering the phalanx of Ancient times. Remember, too, that the Swiss pikemen were superior to the phalanx in that they could successfully charge horsemen, and the rules should allow for this.

9
Sixteenth and Seventeenth Centuries: Start of the Horse-and-Musket Period

IN ABOUT 1300 the cannon came into use in Europe. Used in sieges, it immediately rendered obsolete the medieval castle. Although primitive cannon were used at Crecy in 1346, it was not until about a century later that the French, by discovering the principle of the limber and mounting light guns on carts or waggons, gave cannon the mobility that made them offensive weapons. Soon there arrived on the scene the hand-gun, or arquebus, pioneered by the Spaniards who worked out an effective field tactic for its use. In the Italian battlefields of Bicocca in 1522, and Pavia in 1525, the French pikemen and cavalry charged in vain against the Spanish arquebusiers who lined their trenches and palisades. For some time the development of artillery suffered because commanders preferred the mobility of the arquebusiers. Next came the musket, a longer-range weapon but handicapped in its early stages by its weight, which necessitated supporting it on a forked rest and so made loading and firing a slow process. Arquebusiers and musketeers fought alongside each other until the latter part of the seventeenth century. The invention of the pistol, which was really only a shortened arquebus, later provided the mounted soldier with a convenient firearm.

It soon became obvious that musketeers were easy prey to cavalry, whilst pikemen, lacking a missile weapon, were easily shot down by mounted pistoleers, working just outside the length of the pike. Consequently it was deemed necessary to mix musketeers with pikemen for mutual protection, so causing much military speculation during the sixteenth century as to the proper formation for these mixed companies. The one most commonly used, the 'Spanish' formation, comprised a square of pikemen with 'sleeves' of musketeers at the corners who sheltered under the pikes or entered the squares when the enemy cavalry came to close quarters. Sometimes this had the critical effect of breaking the pikemen's formation and it was not until the invention of the bayonet in the late seventeenth century, making the infantryman a pikeman and a musketeer combined, that a really satisfactory solution was discovered. At first, the bayonet was merely a dagger plugged into the mouth of the musket barrel so that it was impossible to fire the weapon, but later came the ring, or socket, bayonet which fitted round the barrel so that the musket could be fired while the bayonet was fixed.

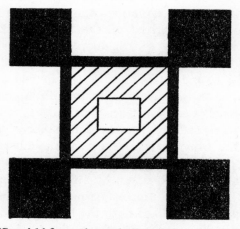

Fig 25   The 'Spanish' formation, a hollow square of pikemen supported by musketeers spread along its sides and in 'bastions' at its corners

The sixteenth century brought another typical twist to military tactics—having rendered the medieval castle obsolete the cannon itself was now defied by up-to-date fortresses and fortified towns. Now began the centuries-long contest of the engineer versus the artilleryman as siege warfare replaced war of movement, so that after Pavia in 1525 there was hardly a pitched battle of any importance for the rest of the century. During this period the Spanish infantry were the best in Europe but strategy and tactics declined, as they always do when the defence is stronger than the attack. It was not until the Thirty Years War (1618 to 1648) that the art of warfare took another decisive step forward when the military reforms of Gustavus Adolphus reached their logical conclusion. Many revolutionary innovations can be credited to this Swedish commander, among them the following.

He entirely remodelled infantry tactics by arming two-thirds of his infantry with a lightened musket that was more simply loaded. Instead of following the current practice of having his musketeers drawn up in ten ranks, firing one rank at a time and then filing off to the rear, so maintaining an unbroken fire, Gustavus formed his musketeers in six ranks which fired three at a time—the first rank kneeling, the second stooping and the

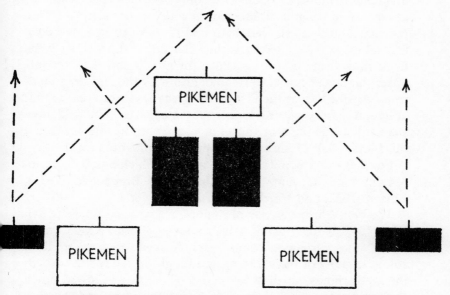

FIG 26 The formation devised by Gustavus Adolphus, made up of mutually-supporting companies of musketeers and pikemen

third standing upright. He also devised a method whereby the six ranks formed themselves into three and together fired a devasting volley.

Gustavus pioneered linear tactics by giving up the solid infantry squares of up to fifty ranks deep and forming companies of 120 men, so offering smaller targets to cannon and allowing more muskets to be brought into action. He drew his infantry up into two long lines, companies of musketeers alternating with companies of pikemen with a reserve behind them. Thrown out at right angles to the line, musketeers were able to use converging fire and, if attacked, they could shelter behind the pikemen, though they were also trained to join in the mêlée with clubbed muskets. Gustavus, seeking reliable and mobile field artillery, introduced the light 3-pdr gun, drawn by one horse or three men, which could be fired as rapidly as a musket; he supplied two to each infantry battalion.

Gustavus also revolutionised the use of cavalry, stopping the manoeuvre known as the 'caracole' in which each rank of cavalry rode up and discharged its pistol at the enemy before retiring. The Swedish leader formed his cavalry up three deep instead of six to ten deep and then charged at the trot, only the front rank firing its pistols. Both for attack and defence purposes, he mixed musketeer companies with the cavalry squadrons. Again, he was the first of the great leaders to realise that although cavalry were no longer supreme on the battlefield, they were still a striking arm and so he insisted on having between a third and a half of his army comprised of mounted men.

For detailed accounts of the manner in which Swedish armies fought, read accounts of the battles of Breitenfeld (1631), Lützen (1632) and Nordlingen in 1634.

The next European war of importance was the English Civil War (1642 to 1651). Although the armies of both King Charles I and the Parliamentarians were formed of inexperienced soldiers, many of their leaders had fought in the recent wars in the Netherlands, so that the tactics of the war followed much the same pattern. Musketeers and pikemen combined together but the Royalist cavalry, under the dashing direction of Prince

Rupert, often charged successfully and then, carried away by their own success, went off in mad pursuit leaving their infantry to their own resources. Apparently, this has always been a common failing of cavalry and one which Wellington suffered with bad grace in the Peninsular War and at Waterloo.

The English Civil War is a fascinating one and too often neglected by those who campaign with model soldiers on table-top battlefields. For details of the manner in which war was fought, read accounts of the battles of Edgehill (1642), Marston Moor (1644) and Naseby (1645), together with the smaller engagements of the war.

The period is much easier to represent in battles with model soldiers than might at first be imagined. The infantry, composed of pikemen and musketeers, must work together for mutual protection, but the pikeman, because of his armour and long cumbersome pike, should be made to move slower than the musketeer. He can act collectively or in a skirmishing role but, although he can move with a loaded musket, he must be stationary to fire. The method was for musketeers to advance rank by rank to fire, the rank which had just fired retiring to the rear to load. This can be represented by musketeers dropping back 3 in when they have fired and moving forward 3 in to fire. Ideally, only part of a regiment should fire, so that there is never a time when *everyone* is unloaded. On their own, pikemen have no fire-power and can defend themselves only with their pikes.

To repel cavalry, a unit of infantry formed of pikemen and musketeers gathered together in a formation known as a 'stand'. In the form of a square, the pikemen presented a hedge of pikes while the musketeers positioned themselves between the pikemen or even under the lengthy weapons. Sometimes the musketeers fired at the advancing cavalry and then rushed back, seeking shelter by flinging themselves on the ground beneath the outstretched pikes. Such a stand could not be broken by cavalry unless the pistols of the cavalry made a breach by killing pikemen. Otherwise the cavalry were held away at pike-length and fired upon by the musketeers sheltering in or under the pikes. But once the horsemen smashed a breach through the pike-

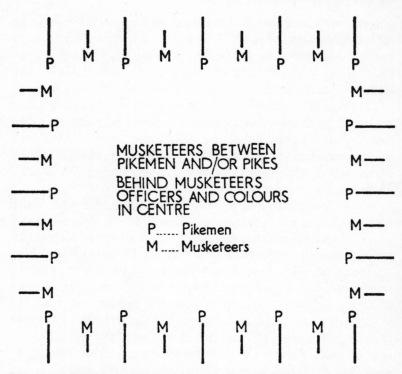

FIG 27   A diagrammatic representation of a 'Stand of Pikes'—the popular defensive formation for infantry when threatened by cavalry during the English Civil War period in the mid-seventeenth century

hedge, the rest of his unit could follow, so rules should be made allowing cavalry to get into the stand as far as their move-distance will permit. When this occurred, the stand was thrown into disorder—so, in a wargame, a mêlée should take place—with the horsemen being given some form of shock-bonus.

In the open, musketeers without the support of pikemen could not stand against cavalry. If attacked, they could fire but if they did not turn the cavalry back by fire-power they broke and fled, and were then cut down by the oncoming horsemen.

Each cavalryman carried two pistols which he could only re-load when he was stationary. The method of attack was for the horseman to trot or walk up to within pistol range of a formation of infantry and then try to force a breach by firing into the pikemen before charging.

It was a period when, as already instanced, cavalry, once given their head in a charge, frequently got out of control as they chased the fleeing enemy, leaving the battle far in their rear. Rules should, therefore, be made to allow cavalry to re-form and be taken under control after a charge only after throwing a reasonably high dice score.

All units, both horse and foot, can vary greatly in their standard of morale and some may fight well to begin with and then, at the slightest discouragement, turn and flee. Sometimes a unit ran at the first shot, only to rally quickly and fight bravely on. Sometimes they would rally but not move forward into the battle, standing as though paralysed with fear despite the frantic urgings of their commanders. A simple way to represent this on the wargames table is to throw a dice for each unit—5 or 6 means that they go forward bravely, 3 or 4 means that they remain motionless, while 1 or 2 means that they flee.

The artillery of the period was slow and cumbersome and rules should be formulated so that the gun takes as much as a complete move to reload. This gives the chance of some very interesting situations when a gun which it is wished to move is unloaded, or when a gun is caught unloaded by oncoming cavalry.

10
Marlburian and Seven Years Wars

BY THE middle of the seventeenth century France had succeeded Spain as the first military power of the Continent. Maintaining a standing army larger than had ever been seen before, France forced other countries to follow suit and this subsequent enlargement of field armies transformed the nature of warfare. Because of their size, these armies were unable to live off the country and it became necessary to establish magazines of ammunition and stores of food at convenient points throughout the country over which the campaign was being fought. Obviously, this restricted the movements of the troops and greatly slowed up the tempo of the war. Simultaneously, the French engineer Vauban and his contemporaries were industriously constructing vast and almost unassailable fortifications around every city and place of importance in the campaign area. These two factors combined to turn wars into a succession of sieges. Armies would not stand and fight unless sure of victory through superiority in numbers or in a strong defensive position; when they did fight, the beaten army could always recover and reform behind its system of fortresses. Thus battles were seldom decisive and generals gained reputations by their skill in avoiding battle and manoeuvring the enemy out of a desired area without actual fighting. It was not until the end of the eighteenth century that this indecisiveness was partially overcome by the military acumen and talents of Marlborough and Frederick the Great.

The Duke of Marlborough won his fame in the War of the Spanish Succession (1701 to 1713). It was a conflict in which France, aided by Spain and Bavaria, fought Britain, Holland, Denmark, Savoy, and many of the states of the German princes.

Besides the war that took place in Flanders, Germany, Italy and
Spain, there were clashes in Asia and America and naval battles
in the Mediterranean.

Field tactics had improved with the virtual disappearance of
the pike, and the substitution of the flint-lock, or fire-lock
musket, for the match-lock had increased the speed and regu-
larity of fire. Realising the importance of firing discipline, Marlborough practised his men in platoon volleying instead of firing
by ranks; their fire was regulated by flag and drum signals. His
infantry formations were less dense and linear tactics prevailed.
Cavalry still continued to form a third or fourth of the total
force, as they were still the only arm that could deal the enemy
a shattering blow. Marlborough discouraged the use of the
pistol or carbine and insisted that the cavalry charged sword in
hand and boot-to-boot, attaching more importance to shock
than to fire-power. In all his battles, the mounted arm played a
decisive part. Little technical progress had been made in artillery
since the time of Gustavus but Marlborough was very much
alive to its continuing value.

For details as to the manner in which these armies fought,
read accounts of the battles of Blenheim (1704), Ramillies
(1706), Oudenarde (1708) and Malplaquet (1709).

Frederick the Great, King of Prussia, inherited in 1740 a fine
army with perhaps the best infantry in Europe. Frederick's ruthless and relentless father had them drilled until they performed
to perfection the complicated evolutions of the time, giving them
a mobility which paid tremendous dividends on the battlefield.
The simple expedient of replacing the wooden ramrod with one
of iron enabled the infantry to shoot three times as fast as other
armies, but their cavalry was so poor that at Frederick's first
battle (Mollwitz in 1741) they ran away, carrying the king with
them, and only the steadiness of the infantry saved the day.
Frederick learned his lesson and before long his Prussian
cavalry were a match for any in Europe. A third of Frederick's
guns were high-shooting howitzers as he was a pioneer in appreciating the value of high-angle fire. But the military operation
most traditionally connected with his name is the 'oblique

order' in which a strong, reinforced wing attacks the enemy whilst the other wing of the army is refused, or held back.

It is said that in only one battle, Leuthen in 1757, did the oblique order function properly. For further details of the methods of fighting in this period, read accounts of the battles of Kolin (1757), Rossbach (1757), Hochkirk (1758), Kunersdorf (1759) and Torgau (1760).

After the Seven Years War, military thinking set about recasting the rigid strategy and tactics of Frederick, and the subsequent introduction of a more mobile method of fighting can be said to have been the forerunner of the tactics of the Napoleonic Wars.

Certain of these innovations stand out because, to a great extent, they revolutionised the warfare of the period. Special battalions of sharpshooters, trained to fight in open order, were organised; the British called them riflemen, the French chasseurs and the Germans jäegers. In addition, all ordinary infantry battalions had a light company which was thrown out in front of the attacking or defending line. The need for such tactics had been revealed by British experiences in North America, when their stiff lines of regular troops were helpless against loose formations of Indians and Colonists firing expertly from behind cover.

One great military argument which raged throughout the Napoleonic Wars concerned the respective merits of line formations against columns. By bringing the greatest possible number of muskets into action, line formation gave superiority of fire but sacrificed a certain amount of shock power, besides being liable to be broken by superior weight and depth on the part of the enemy. In addition, a line of troops moving across broken ground could be thrown into disorder because of difficulties of control. The French Revolutionary armies, lacking in discipline and training, were considered easier to control when moving in close-ranked columns surrounded by crowds of skirmishers. First the French used battalion columns, then brigades, and later Napoleon had columns of divisions—the solid blocks of troops which hurled themselves against the British at Waterloo.

Few armies could withstand the assault of the French columns, but in the Peninsular War the British veterans, in lines two deep with a screen of skirmishers in front and, wherever possible, sheltered behind the crest of a ridge, invariably defeated the French troops in column formation. As the columns approached, the skirmishers would run back to announce their approach, and when the huge formation appeared over the ridge it was blasted with musketry fire. Then, as it reeled from the volleys, it was charged by the British line whose wings folded in on the flanks of the column.

The splitting-up of armies into divisions composed of infantry, cavalry and artillery left them free to manoeuvre independently, each as a small army. Previously, when armies fought in single blocks, there was little point in interposing between an enemy and his base because invariably he could find a way round. The divisional principle enabled an army to envelop the enemy by scattering its forces over the countryside to cover threatened points or, if the enemy attacked, to concentrate rapidly and to hold him with a few divisions until the remainder came up. Lines of communication had a new significance, to get across them was to cut off an army's supply of food and ammunition and manpower replacement, and an army so threatened was forced to fight. It is not unreasonable to claim that the divisional principle was the most fundamental change in military practice of the eighteenth century.

The infantry of Marlborough's campaigns and Seven Years War fought in large, tightly-packed formations. Harshly disciplined so that they obeyed orders without question and kept in position by officers and sergeants, the men moved ponderously as one, loading and firing massed volleys of musketry in oft-repeated drills controlled by shouted orders. They fought in this fashion partly because they were rigidly drilled to do so and partly because their commanders dare not let them fight in looser formations for fear of desertion or retreat.

Thus it can be seen that rules governing battles in this period should tend to allow infantry to move in a formal pattern where firing, moving and facing in a different direction all take up a part of their move so that a choice has to be made as to which two of the three actions are performed.

Effective musket range should be kept short so that only the closest-range firing has very much effect. Artillery was often brought on to the battlefield by civilian drivers who, before they hastily retreated to a place of safety, positioned the guns where they were to remain for the rest of the battle. At anything except fairly close range, these guns were not particularly effective as the round iron ball they fired was not a very tight fit in the un-rifled barrel and the recoil of the guns made it necessary to re-lay them after each round.

Because of the close order in which they fought, the rigid discipline to which they were accustomed, and through fear of the officers and sergeants who hemmed them in so closely, these infantry rarely suffered any failure of morale during the battle. They were professional soldiers who fought where they were placed and died there if necessary. When, after suffering in-credible losses, they broke and fled it was for good, there was no rallying them. For these reasons the wargamer can see that any morale rules he formulates must differ from those for other periods and must only affect his units after the severest reverses, and should then make it almost impossible for the units to rally and return to the battle.

In the latter half of the century, advanced military thought reasoned that all men were not likely to desert and that well-trained soldiers could be imbued with sufficient pride and initia-tive to form valuable irregular or light infantry. In our table-top battles, such men can split-move—that is to say they can move, fire and move again, or fire and then move. Their movements can be far less formal and cover more ground than the ordinary infantryman and they will not fire in volleys but as individuals.

Each regiment had a company of grenadiers, frequently the biggest and strongest men in the unit, who originally threw grenades but later became the élite, or shock, troops who always formed up on the right of the regiment and, because of their higher morale and ability, were used to spearhead attacks. In table-top battles, the grenadier company can be given additional powers so that they correspond to guards units of other eras. Frequently the light, or grenadier company of each regiment would be withdrawn and banded together to form one élite regiment of light troops or grenadiers for some specific operation.

The eighteenth century is an interesting period in which to fight battles with model soldiers because it was both colourful and set a new pattern in the use of massed formations with a formal style of moving. Also, it does not involve those disastrous withdrawals through faulty morale which can be so disconcerting to the wargamer—in this period he may pretty well rely upon his troops remaining firm until the final stages of a battle when their numbers have been greatly diminished.

11
Napoleonic Campaigns

NAPOLEON'S SUCCESSFUL principles of war can be summarised under four main headings.

1. *Superior Numbers*. Always endeavour to be stronger than the enemy. Superior strategy and tactics do not compensate for inferior numbers. In this connection it is worth pointing out that, by themselves, strategic moves never win campaigns as they contain the built-in paradox that although a great strategic victory, such as cutting the enemy's line of communication in his rear, could destroy him, if he defeats you and places himself between you and your base, then all the consequences you prepared for him fall upon yourself.

2. *Concentration*. It is necessary to concentrate every available man on the battlefield to secure superiority of numbers. Although an army may be strung out for some initial strategy, it must be capable of being rapidly concentrated when the moment of battle arrives. Split forces are disastrous as they can always be beaten one by one by the superior numbers flung against them.

3. *Unity*. This is closely allied to the previous principle in that divided commands and duplicated fields of operations lead to dispersion and weakness. It is essential that in each field of war there should be only one commander and his army must have a single objective.

4. *Speed*. Battles are won by being quicker than the enemy, by seizing the initiative and keeping it. Speed of execution will frequently enable a general to beat an enemy who has a better battle plan than himself.

It has been said that Napoleon's strategy in his early Italian campaigns was never improved upon and that all that followed was mere repetition. The lesson of which would seem to be that if you have a good plan then stick to it!

Napoleon's tactical recipe for victory was '. . . it is by turning the enemy by attacking his flank, that battles are won'. In practically all his battles, not always successfully, Napoleon practised the relatively stereotyped manoeuvre of attempting to get on his opponent's flank or rear. He either tried to do so by bending forward one end of his line to overlap the enemy or, preferably, by detaching a force to come round behind him. Keeping the enemy busy by demonstrating against his front, Napoleon waited until the guns of the flanking force announced they were in position, then he pressed his frontal attack and threw in his reserves.

Napoleon formed his infantry in columns in such a way as to obtain a combination of fire-power and shock. In his earlier battles he used a mixture of line and column, known as the 'half-deep' order, in which one battalion deployed in line with a battalion in each flank in column in double companies. It was a formation which could be extended indefinitely with line and column alternating (see Fig 28). Later, Napoleon changed to solid columns without mixing line so that they formed an infantry mass. He used cavalry in orthodox fashion to press pursuits and to finish off the enemy and sometimes to assist hard-pressed infantry, as at Eylau and Waterloo. Believing in massed artillery fire, Napoleon massed batteries to hit vulnerable points in the enemy's lines and soften them up for infantry assault. In his later career, when he had as high a proportion as

FIG 28   The French (Napoleonic) 'Half-Deep' Order

three or four guns per thousand men, he surprisingly substituted these tactics for manoeuvring.

The Napoleonic Wars are easily the most popular to recreate in battles with model soldiers. There is much to recommend in this era—many nations were involved in numerous parts of the world so that the wargamer has a wide choice of different soldiers and uniforms; vast amounts of literature and information are readily obtainable and the range of model soldiers is extensive. Rules are not difficult to formulate so as to obtain a realistic simulation of this period, providing certain basic factors are considered.

Infantry still fought in close formation, although light troops such as the famous British Light Division and the French skirmishers played important roles. Accounts of battles of the period tell us that the French frequently attacked in dense column formation with frightening effects upon the infantry of all nations except the British. Under the cool leadership of Wellington, the British infantry refused to be panicked by the alarming sight of a dense column moving steadily towards them preceded by the demoralising beat of dozens of drums. Forming up in double or treble line, the British infantry were able to bring to bear musketry fire from the entire regiment in a shattering volley. Before the temporarily demoralised formation could recover itself, the British charged in with the bayonet, the flanks of their line overlapping the sides of the column.

To recreate this realistically on the table-top battlefield it is necessary to remember that the steadfastness of the British infantry was due to their exceptionally high standard of morale. Thus, when formulating rules, the British infantry should be given a superior morale-rating over those of other nations involved. One reasonable method of tackling this tricky problem of representing a column attack is as follows:

First, a column must be a column with a minimum size of four men wide and five men deep, extending upwards to any desired width or consisting of any number of men. A column may fire when approaching, although the only men who can fire are those in the front rank and those on the outside of the files.

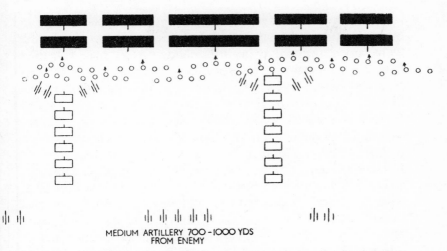

MEDIUM ARTILLERY 7OO-IOOO YDS
FROM ENEMY

Fɪɢ 29 A schematic representation of French battle tactics during the Napoleonic period. First: the French tirailleurs engage all along the enemy front supported by medium artillery. Second: French infantry columns, accompanied by light artillery, come forward to attack one or more points in the enemy line. These points may or may not have already received concentrated fire from medium artillery. Third: light artillery go into action at ranges of 150–300 yd against the enemy line at the points where the columns will strike. The aimed fire from the tirailleurs, the light artillery close-range fire, plus possible medium artillery fire, and finally the threat of infantry bayonet shock, usually caused the enemy to break before contact. Reproduced from *Weapons and Tactics: Hastings to Berlin* by kind permission of the author, Jac Weller, and the publishers, Nicholas Vane

When a column comes within charging distance of a line of infantry other than British, the defenders must throw a dice to see if they stand firm, requiring a score of 4, 5 or 6 to do so. If they do not stand then they retreat 12 in directly backwards. If a column loses one-third or more of its strength during its approach, then it also requires to throw a dice to see whether it charges home. If it does not make the required score of 4, 5 or 6, then it turns halfway towards the infantry and retreats back to its starting point. When attacking a British line, a column will need to throw a dice to see if it charges home when it receives *any* fire.

When the column reaches the enemy line it penetrates 3 in into it, pushing before it the same number of men as occupy the front rank of the column (ie, if the column is four men wide then it pushes four men back before it). The remainder of the defender's line curves back to conform so that the flanks of the line remain in their original position while the centre of the line is curved back 3 in. A mêlée then takes place between the front two ranks of the column and the men it has pushed before it, with the column adding 1 for each dice it throws in the mêlée. If a second round of mêlée fighting is necessary, those defenders who are in contact with the flanks of the column also take part, with the column flank men also adding 1 to each of their dice thrown in the mêlée. Unless the defenders push the column back after the first round of fighting, then the column gains a further 3 in during the next round of the mêlée and a similar distance for each round that the mêlée lasts.

If the British line carries on in its accepted fashion (firing a volley and then charging home) then, instead of the column being the aggressors, the British have snatched that role from them. This is represented by removing from the column the men killed by the volley and then moving the British forward to make contact. In the resulting mêlée, the column scores as usual in mêlées, whilst the British add 1 to each of the dice they throw in deciding their mêlée score.

Another aspect of Napoleonic warfare which must be reflected if realism is to be obtained was the infantry units' ability to repel cavalry. In warfare of this and earlier periods, the battlefield was soon shrouded in a murky mist generated by repeated discharges of small arms and cannon plus, in some cases, the dust that arose from the movements of thousands of men. With visibility down to a few yards, it was possible for swiftly moving cavalry to pounce upon infantry before they could organise themselves to handle the threat. Because of the short effective range of his musket, an infantryman was unable to bring down enough of the onrushing cavalry to halt their charge and the crushing impact of horse and rider smashing into a single or double line of infantrymen was more than the foot

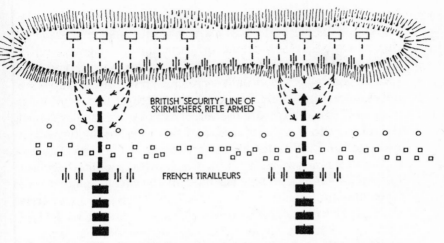

FIG 30 Wellington's Defensive Position. British battalion columns at quarter distance on the reverse slope. On forward slope or crest British light artillery are positioned. This schematic representation of Wellington's counter-tactics is reproduced from *Weapons and Tactics: Hastings to Berlin* by kind permission of the author, Jac Weller, and the publishers Nicholas Vane

soldier could withstand. Consequently, the infantrymen had to form up into a tight cohesive formation to repel cavalry, a simple revival of an ancient practice. Just as the Macedonians and the Greeks had formed a phalanx, and pikemen and musketeers had gathered together in their 'stand', so the infantrymen of the Napoleonic Wars formed a square. With officers and colours in its centre, the square consisted of three or four ranks, the first kneeling with rifle butts on the ground so as to present a hedge of glistening bayonets whilst all ranks alternated in firing and loading. So rare were the instances of cavalry forcing their way into these squares that they are well-known and recorded. Perhaps the best-known instance of the employment of squares during the Napoleonic Wars took place at Waterloo on 18 June 1815, when the British square stood firm for hour after hour whilst surrounded by milling clouds of French light and heavy cavalry.

Another reason for the formation of these infantry squares

lends itself ideally to a tactic on the table-top battlefield. The musket of the period had an extremely short effective range so that it was easily out-distanced by cannon firing grapeshot (a short wooden rod with a wooden disc at the base served as a core around which iron balls were clustered, with a grape-like appearance, the whole being bagged in cloth and reinforced with a net of heavy cord). The threat of hovering cavalry forced the infantry to form a square, and this closely-knit formation presented an ideal target for the grapeshot of the guns. The unfortunate infantryman's position was, however, counterbalanced to some extent by the immense difficulty the cavalry encountered when attempting to storm a square. Though, in fact, they rarely did manage to break in, a wargame's rules should be so devised as to give them a slight chance of success without unfairly penalising the infantry, who have already had more than their share of bad luck from the guns! A reasonable and not too involved method of representing this situation is as follows.

When approaching, the attacking cavalry take musketry fire when they come within range of that face of the square which they are attacking. Because of the confidence engendered by their tight formation and the feeling that they can give something back, the infantry might possibly fire more steadily than usual, so it is not unreasonable to give them some sort of a bonus to their firing score at this stage. Should the cavalry lose *a quarter* of their numbers during this approach, then they must throw a dice to see whether they charge home—1, 2, 3 or 4 means that they turn back, while 5 or 6 means that the cavalry force split and continue the rest of their move along either side of the square, taking musketry fire from those faces of the square which they pass during their passage. Should the cavalry not lose a quarter, then they throw a dice again and need a score of 5 or 6 to charge home, when a mêlée will take place with the infantry counting double for their cohesion. If the cavalry have not broken into the square after one round of fighting, then they are considered to have been thrown back and they turn and retreat.

Also requiring consideration is the different types of ammunition used by the artillery. Grapeshot has already been considered; canister was similar to grapeshot except that the balls were smaller and contained in a can, spreading out as they left the muzzle of the gun. The British used exploding shells but, in the main, the guns fired solid iron shot of varying weights. These balls cut bloody lanes through the massed formations of infantrymen and then bounced on across the battlefield in decreasing bounds, devastating all in their path. In other words, a projectile from a gun did not hit only the first thing it reached and a system has to be evolved which allows for this bouncing effect and the casualties a ball could cause behind the foremost lines of troops.

The Napoleonic period, because it is a particularly interesting one in which to battle with model soldiers, is also the period for which the wargamer is most likely to be able to find an opponent.

12
The Crimean, American Civil and Franco-Prussian Wars

THE CRIMEAN War and the Franco-Austrian War of the mid-nineteenth century were both fought in much the same way as the Napoleonic Wars. In the Austro-Prussian and Franco-Prussian Wars, infantry still attacked in dense lines and cavalry still attempted dashing charges, while as late as 1877 the Russians at Plevna advanced in solid columns against Turks who were armed with repeating rifles. All this happened because the notoriously conservative military mind required an unreasonable amount of time to realise the extent to which improvements in weapons had revolutionised tactics.

By rifling their barrels, the range of smooth-bore cannon was increased from 1,000 to 4,000 yd, and these rifled cannon came into use about the middle of the nineteenth century. In 1841, the Prussian army were equipped with the first breech-loading rifle, followed twenty years later by magazine or repeating rifles. Smooth-bore muskets, such as the British 'Brown Bess' used during the Napoleonic Wars, had an effective range of less than 100 yd—the new weapons had an effective range of as much as 600 yd. Another artillery improvement was the replacing of the solid cannon-ball with shells which exploded on impact or by means of a time fuse.

These mid-nineteenth century innovations introduced changes in warfare which were not fully appreciated even more than half a century later, in World War I. Now it was only possible for offensives to succeed when they possessed overwhelming numerical superiority combined with ability to execute flanking or turning movements. The defence possessed an enormous advantage as a result of the longer range and greater accuracy of the new weapons, so that the only possible attack formation was a loose line of skirmishers who, even so, suffered heavily in frontal assaults. Against even a single line of riflemen, cavalry were helpless and, ceasing to be a striking arm, were reduced to acting as scouts.

And yet this new phase in tactics—there for all to see in the American Civil War of 1861–5—was totally ignored by the professional soldiers of Europe because that war had been fought by amateurs.

Few wars can be of greater interest or offer more opportunities to the man who campaigns with model soldiers than the American Civil War, and most wargamers at some time or another reproduce battles between Federals and Confederates on their table-top terrain. It was a war that had everything for the 'horse-and-musket' enthusiast, and one that combined campaigns and battles with units, both large and small, which were colourful in appearance and bold in action. Cavalry charged in the normal fashion, made raids deep into enemy territory, and fought dismounted with rapid-firing carbines. The rifles of the infantrymen out-ranged grape-shot with shattering effect so that artillerymen had to work under heavy fire; there were sieges and amphibious operations, and even naval battles involving not only wooden ships but the first of the iron-clads. Trains and railways, too, played a big part in the campaign and enable the wargamer to introduce some highly ingenious new factors into his table-top battles. Certainly the American Civil War is well worth study by anyone who wishes to fight really interesting table-top campaigns and battles.

Apart from the Victorian colonial campaigns, the Franco-Prussian War of 1870–1 is the last of what, for wargaming purposes, can be classified as 'horse-and-musket' games. The French lost this war on all counts, in spite of their infantry being equipped with the chassepot rifle which considerably out-ranged the Prussian infantry's needle-gun, and even though they had batteries of an early type of machine-gun, the mitrailleuse.

A few years previously, in the war against Austria, the Prussians had suffered heavy casualties when their smooth-bore cannon were hopelessly outranged by the Austrian rifled cannon. This had been remedied by 1870 and the quality of the Prussian artillery and the manner in which they were used played a big part in their victory.

For details of the manner in which the Austro-Prussian War was fought, read accounts of the Battle of Sadowa (1866). The principal battles of the Franco-Prussian War were Worth, Spicheren, Gravelotte, Mars-la-Tour, Sedan, and the siege of Paris.

This is a most interesting period for the wargamer because developments in weapons had led to improved tactics which completely altered the complexion of 'horse-and-musket' warfare as fought in the Napoleonic periods. Its difference from those periods was as great as that which marked the early days of gunpowder and firearms, and it is important that the wargamer should grasp this fact if the rules he formulates are to give a realistic representation of battles of the period.

It will be recalled that, in the Napoleonic period, infantry were forced into squares by the threat of cavalry attack. Tightly packed together in a defensive formation they dare not abandon, the unfortunate infantry were shattered by blasts of grapeshot from cannon beyond the effective range of their muskets. The battles were fought in a choking fog of powder-smoke and dust, making it possible at any time for a regiment to be surprised by a sudden rush of the enemy from out of the fog. The infantry were forced to remain standing if they wished to continue firing because it was only possible to load muskets in a standing position.

All this had altered by the second half of the nineteenth century. The old muzzle-loading muskets had gone and the infantryman was now armed with a greatly improved rifled weapon which brought the crews of guns within his range, so that he no longer had to stand helpless facing a hail of grapeshot. If a gunner wanted to fire his most effective projectiles at point-blank range, he had now to be prepared to take volleys of return fire from the infantrymen. Another revolutionary aspect of the new firearms was that they were breech-loading and could be reloaded with the firer lying down in comparative safety. This put an end to the massed, standing formations that presented such wonderful targets to each other. This increased fire-power and range now made the cavalryman's life far more hazardous, and military history records numerous instances of large-scale cavalry charges turned back by sheer weight of fire-power before getting anywhere near their infantry target. Finally, towards the end of this period, the guns and muskets used a smokeless powder so that the battlefield was no longer shrouded in a murky and dangerous fog.

A set of simple and reasonably realistic rules has already been used in this book to demonstrate a battle of the American Civil War period. Obviously, they represent basic levels of war-gaming and, aided by what he has already read in these pages, the intelligent wargamer will soon be able to include additions and refinements based on what actually occurred during the 'horse-and-musket' period.

To indicate the direction in which such refinements can lie, let us consider one of the peculiarities of the American Civil War in so far as it applies to the rules governing battles with model soldiers. It was a conflict in which the armies of both sides were formed of independent men unaccustomed to the rigid discipline of European armies. This attitude encouraged personal initiative and considerable fluidity of action, but also resulted in a marked lack of discipline and a tendency to question orders and deci-sions which were not immediately clear or acceptable. On the battlefield this was reflected by regiments momentarily panick-ing and breaking, only quickly to reform and fight on as bravely as though previously unaffected. To reproduce this factor when battling with model soldiers it is necessary to have morale rules that allow such breaks to be transitional and for the units to rally more easily than their European counterparts. On the wargames table this can be reproduced in the following manner:

On the FIRST occasion a unit comes under fire it has to be dis-covered whether it stands firm or breaks. A dice is thrown:

4, 5 or 6—the unit stands firm.

2 or 3   —the unit breaks and retires in good order for half its normal move-distance and ends up facing the enemy.

1   —the unit turns and flees 12 in to end up with its back to the enemy.

Coupled with the last sub-paragraph is the necessity for a rule enabling a unit to be rallied at the start of the next move, when its morale state can be heightened by the presence of a general or by the commander-in-chief of the entire army (reflected by adding 1 or 2 to the score of the dice thrown to decide the unit's morale-state).

13
Colonial Wars

DURING THE eighteenth century the soldiers of Britain, France, Spain and Holland fought each other, seeking to acquire African or Indian colonies for their monarchs. Sometimes they fought the fierce but ill-disciplined natives of the area, on other occasions they were pitted against well-drilled armies of native soldiers from the states of the Indian princes. Once conquered, many of these native armies were enrolled to serve as regiments under white officers and fought alongside unsuitably uniformed European soldiers. On the American continent, both French and British enlisted the native Indians to assist them in scouting and fighting. In addition, they had bands of white men born in the country, such as Roger's Rangers.

In the campaigns of Clive in India, small numbers of British troops with Indian sepoy battalions defeated larger numbers of native troops backed by French regiments at such battles as Plassey and Arcot. This is a little fought but wonderfully suitable type of warfare for reproducing on the table-top. The French and Indian wars of the mid- and late-eighteenth century, together with the capture of Quebec in 1759, represent another era fruitful in its opportunities for reproducing warfare with many differing types of troops, to say nothing of small-scale naval engagements on the lakes.

It was not until the nineteenth century was well under way that these colonial campaigns really began to hot up as the major powers realised that time was running out and that there were not many native territories, rich in minerals and resources, left to conquer. But the complexion of these colonial wars had altered, changing from wars between Western powers and their native allies into wars between the Western powers and the savage tribes and races still clinging tenaciously to territories which had previously been considered unsuitable for conquest. In other words, the Western powers had tacitly agreed upon their areas of influence and were now consolidating their holds.

The principal pioneers in these colonial wars were the British and the French, but while the French restricted their activities to North Africa and, later in the century to Indo-China, the British, with their native allies, marched steadily forward under a succession of commanders all of whom seemed to be cast in the same mould and all ideally suited to the task of adding to Queen Victoria's realms. Their activities took them to many countries and, almost without a reverse, victories were won in Abyssinia, Afghanistan, Ashanti, Australia, West Africa, Burma, Canada, China, Egypt, the North-West Frontier of India, Japan, the Levant, Malay, Mashonaland, Mombasa, Natal, New Zealand, Niger, Persia, Somaliland, South Africa, Sudan, Tibet, Transvaal and Zanzibar. In these countries they met and defeated Africans, Asians, Dervishes, Maoris, Sikhs, Zulus, Kaffirs, Afridis, Pathans, Mahrattas, Baluchis, Afghans, and even rebellious Canadian half-breeds.

Invariably, the British soldiers were out-numbered but made up for this by superior tactics and far better weapons. But victory was not entirely due to the mountain gun and the Martini-Henry rifle. As much as to any other factor, success was due to the innate and cheerful courage of the uncouth and harshly disciplined British soldier who realised that, in these desperate affairs, the difference between outright success and stalemate was to be wiped out.

Even the British Navy took a hand as landing parties of straw-hatted sailors and marines landed from men-of-war to accompany the expeditions. Sometimes they dragged Gatling and Gardner guns through the hot sands of the Sudan or the tough scrubby grass of Zululand. Peel's sailors from the *Shannon* performed wonders with their heavy guns at Lucknow and Cawnpore during the Indian Mutiny, whilst 4·7-in naval guns and their crews did sterling service in the Transvaal during the Second Boer War.

Probably the greatest training ground the British Army ever had was the North-West Frontier of India. Here, for almost a hundred years, practically every regiment in the British Army, together with such reliable allies as the Guides, the Gurkhas and other well-known native regiments, fought against the redoubtable hillmen on their home ground.

Other enemies, lacking weapons with the range of those used by the British, relied on incredible courage to attack in fast-moving masses which broke like waves on the rock of British resistance. The most formidable opponents with this style of fighting were the Zulus of Africa and the Dervishes (Fuzzy-Wuzzies) of the Sudan. The former were a warrior nation whose men were grouped in distinct regiments which carried out well-conceived tactical manoeuvres with astonishing speed and a complete disregard for death. The Fuzzy-Wuzzies, together with the Maoris of New Zealand, earned an almost affectionate respect from the British soldier who knew bravery when he saw it. So fierce were their onrushes that rarely did one bullet halt a man and, as Kipling so ably commemorated, they '. . . broke a British square' (Abu Klea, 1885).

Not all Britain's native opponents were savages, nor did all of them rely only on sheer courage. The Afghanistan forces in the second and third Afghan Wars were well-drilled and uniformed native regiments with modern European artillery and gunners trained by Russian officers and military missions. At battles such as Maiwand in 1880, the Afghan artillery was not only of a better quality than the British but far more numerous. Similarly, the first and second Sikh Wars of 1845 and 1848 brought the sepoys of the British East India Company into conflict with a warrior nation trained by European mercenary officers (many with experience of the Napoleonic campaigns) and equipped with what Lord Hardinge, Governor-General of India, described at the time as the finest artillery in the world. The two Sikh Wars form ideal models of conflicts which can be reproduced on the table-top battlefield—the native enemy wore colourful uniforms, the British regiments were still dressed in red coats and fought exactly as they had done in the Peninsula forty years earlier. The Battle of Aliwal in 1846 was considered by Sir John Fortescue, the historian of the British Army, to be 'the almost perfect battle'.

Two wars against the Boers of the Transvaal in 1881 and 1899 possessed many unique features and enable the wargamer to reproduce in miniature the fascinating but difficult type of warfare in which fast-moving irregular troops, fighting on their own terrain, hold off the more numerous and better equipped but slower moving British regular troops.

Whilst all this was going on, the French were colonising North Africa, using at first the regular French Army, and then the famed Foreign Legion together with such colourful regiments as the Zouaves, Turcos and Chasseurs D'Afrique. Under the hot sun of an arid, desert country, the French for more than a century fought against brave and wily Arabs. Small isolated forts were defended against large bodies of natives, and Arab towns such as Constantine were taken by storm. Later in the century, the Foreign Legion and another excellent French colonial force, the Marine Fusiliers, fought some very hard campaigns in Cochin and Indo-China. All in all, a colourful and interesting but sadly neglected period for the wargamer who wants an unusual table-top campaign.

At the very end of the nineteenth century, the Germans were engaged in South-West Africa and in Togoland, striving to grab a little of what was left in the way of colonial possessions. They enrolled battalions of native troops, known as Askaris, and their battles followed the usual course when Western nations opposed natives on their own ground.

At about the same time, the Italians, in 1896, were taking a beating from the Abyssinians at Adowa, but recovered sufficiently later to colonise Libya. The Spanish, too, had their Foreign Legion and their troubles with Riffs in North Africa, while the Americans beat the Spaniards in a semi-colonial war in Cuba, a campaign more marked by death from disease than in battle.

Just as the century ended, an incredible conflict took place in

China when a patriotic military force known as the Boxers be-
sieged the foreign legations at Peking. A relief force was hastily
organised, formed of soldiers and marines from all the nations
whose representatives were in danger. Under the command of a
German, it consisted of British, Americans, Austrians, Japanese,
Indians, French, Italians, Russians and Germans. After some
hard-fought battles, the force reached Peking and relieved the
beleaguered international garrison, which had manfully held
out with military and civilians alike manning a strange assortment of weapons.

So far as battles with model soldiers on the table-top are concerned, it can readily be seen that this is not a single period or
type of warfare but a mixed collection stretching over a period
of nearly 200 years, with the emphasis on the last sixty years or
so. The battles against savages such as Fuzzy-Wuzzies and
Zulus will need different types of rules to the battles fought
against disciplined native troops, such as Sikhs and Afghans.
The latter type of battle can be fought much to the same rules as
are used for 'horse-and-musket' battles of the same period. It
will be necessary to make adjustments, basically in the morale
section of the rules, to compensate British inferiority in numbers
for their greater fighting spirit and courage. On the other hand,
the fighting ability and courage of the native enemy should not
be underrated. The Sikhs and Afghans, for instance, were
armed as well, if not better than their opponents, so that the
difference in fire-power, except in so far as it is affected by the
steadiness (ie, morale), would not greatly differ.

Perhaps the simplest method of handling these differences is
to make a difference of 1 in the score of dice thrown to indicate
the morale-rating of either side. Thus, any dice thrown for this
purpose to cover a British group would add 1 or, conversely,
leave the score as it is but deduct 1 from any dice score of the
native enemy. A similar situation could prevail when considering the morale of native troops fighting with the British, so that
they benefit from the immediate support of white regiments and
conform to any detrimental moves by an adjacent European
unit. This means that if a white regiment is forced to fall back,

then any native regiments on its flanks would conform to that movement and fall back with them. The first thought that springs to mind is that this can be considered a trifle unjust on regiments with such formidable reputations as Gurkhas and Guides, for example. One way of adjusting this problem is to co-ordinate the morale of native units with the loss or otherwise of their white officers. Thus, a native infantry regiment which loses one of its officers during any specific move is considered to be in a far more shaky position than a similar regiment with all its officers remaining.

If battles between large numbers of brave but ill-disciplined natives and white troops with their native allies are to be reproduced satisfactorily, a careful balance in the rules is necessary. In the first place, the strength of the natives lay in the surprise element of their attack and the shock effect of their fearless headlong charge. Surprise is immensely difficult to simulate in battles with model soldiers and the on-rushing attack by large numbers of warriors would, unless the rules are suitably adapted, merely result in their being turned back by the superior fire-power of the formed troops before they came anywhere near mêlée range. Indeed, this was frequently the case in real life but it does not make for a particularly interesting battle on the table-top. One way to balance-up the situation is to make the attacking native force considerably stronger than the formed troops—this measure in itself can sometimes make all the difference that is required. Another, and possibly an improvement, is to increase the mobility of the natives (ie, give them a longer move-distance than the formed troops). Some allowance should also be made for surprise, so that the formed troops are not able to get off the full number of volleys to which they would normally be entitled.

When the natives actually get to grips, they should be given increased powers in the resulting hand-to-hand fighting. For example, a big Fuzzy-Wuzzy who has spent a life of considerable hardship in the desert and is an expert with his razor-sharp sword or broad-bladed spear can reasonably be considered more than a match in hand-to-hand fighting for an average-sized British soldier who probably only joined the Army because he was hungry and unemployed and is now thousands of miles from home in an unsuitable uniform under a burning sun faced with vastly superior numbers of terrifying natives. So it is reasonable, in the melée, to give the native a value of $1\frac{1}{2}$ points and the soldier a value of one point. Practically reflected, this means that ten soldiers would be worth ten points fighting ten natives who would be worth fifteen points—so that the natives throw three dice to the soldier's two (one dice being thrown for every five points).

The natives should also be given a generous shock-bonus when they make contact. The headlong charge of a brawny savage wielding a huge spear is worth 2 added to each dice thrown for his group in the melée.

Like the feudal troops of centuries earlier, natives were led by their own chieftains and leaders, and rules should be adjusted so that they derive some sort of benefit when so led; equally, the loss of a chief can be regarded as having a demoralising effect on them. Natives often fought very bravely so long as their chief was unharmed but on many occasions a battle going well for natives was lost because the chief was killed or severely wounded. When natives led by a chief or leader attack in a melée, it would be reasonable to stipulate that no dice they throw should be allowed to score less than 3.

If, in the first round of melée fighting, the natives kill more men than they lose themselves, they can then be considered to have broken into the troop's formation. This would result in a great milling mass of fighting men so that ALL men in the melée within a certain distance (say 6 in or 9 in from the centre of the melée) count instead of only the men in the front. Fighting will go on for as long as one or other of the sides stands.

The undisciplined but brave impetuosity of the natives can be reflected when the formed troops break and start retreating, then the natives will impetuously chase them a further 6 in. If they have a chief with them, they may escape this forward rush by throwing 4, 5 or 6. Such a move forward may mean that the melée may carry on for the next move.

Having weighted things heavily in favour of the natives, it is also necessary to give the formed troops some realistic benefits to ensure that an interesting game results.

If the attacking natives lose more than 25 per cent of their numbers in the approach-charge, they then need to throw a dice to see if they can carry on and charge home. If they score 4, 5 or 6, then they go into the mêlée, otherwise they fall back to their starting point and will require a 4, 5 or 6 throw at the beginning of the next move in order to advance again.

If the formed troops are not less than a third of the strength of the attacking natives, but still hold their ground after two moves of fighting, then the natives must retreat their full move-distance and may not renew the mêlée during the next move.

Bodies of natives attacked by formed troops will not withstand such attacks unless they can throw a dice score of 4, 5 or 6. Native cavalry will similarly need to see if they stand when attacked by disciplined cavalry. No natives will stand if attacked by cavalry bearing lances, as natives had a wholesome respect for the lancer with his long weapon with which he could reach out and catch them as they ran or lay on the ground attempting to hamstring his horse. Natives so attacked will break at once and can be ridden down by the cavalry.

Many colonial campaigns were fought in jungles and under conditions of great territorial disadvantage to the British troops. Ambushes were frequent and formed one of the strongest weapons the natives, fighting on their own terrain, could offer. Jungle fighting and ambushes are far from easy to simulate in table-top battles with model soldiers and it might be best if such intricacies were left until the novice has mastered the basic fundamentals and progressed to a more advanced stage of wargaming.

14
World War I

WORLD WAR I (1914–18), in its major theatre of France and Flanders, sprawled from Switzerland to the sea so that neither side had any flanks to turn. So huge were the contesting armies that movement was impossible and strategy became a lost art. Lacking flanks, only frontal attacks could be made, but so strong were the defences and so supreme the machine-gun that break-throughs were almost impossible and most of the fighting from the end of 1914 to November 1918 consisted of indecisive trench warfare. In such a situation, where neither side could find the necessary inspiration to alter the state of affairs, only the British invention of the tank indicated a step in the right direction, though it was quickly ruined by unintelligent handling.

In the open warfare of the first three months of the war, the French and British armies managed to halt the German march before it reached Paris. Then followed the 'race to the sea' as each army tried to outflank the other and, as winter closed in, ribbons of trenches scarred the landscape from Switzerland to the sea and there began the deadlock that was to grip the Western Front for three years. Up to then, the Germans had advanced almost recklessly in much the same manner as they did in 1870 whilst, in other parts of France, the French, regardless of the numbers, armament and position of the enemy, followed a blind policy of attack that cost them more than half a million casualties in three weeks.

Nevertheless, this open warfare of 1914 is about the only period of World War I on the Western Front that lends itself to campaigning with model soldiers. The French, still wearing their baggy red trousers and with cuirassiers adorned with brass helmets and gleaming breast-plates, can fight against Prussians in pickelhaubes, while the khaki-clad, flat-capped British Expeditionary Force fought dogged actions at Mons and Le Cateau. After that, trenches and barbed wire, backed by innumerable machine-guns, make table-top battles a rather dull affair, although there is considerable scope for the small-scale trench raid as one side or the other attempts to secure information from the prisoners they bring back.

The first possible answer to the problem of breaking through the strongly defended areas came at Ypres in April 1915, when the German use of poison gas tore a huge gap in the French front. As always, defence rallied to overcome attack and gas respirators negatived any further such surprise for the remainder of the war, although both sides continued to use gas, chiefly in the form of gas shells. The Allies placed great reliance on the artillery barrage as an offensive weapon, believing that guns could smash up the enemy trenches, kill off the defenders and beat flat the maze of barbed wire that fronted them. Their faith was misplaced as artillery bombardments seldom destroyed the whole of a trench garrison and survivors, emerging from deep dug-outs, would quickly man their machine-guns and wipe out the oncoming infantry. About the only British offensive that showed any promise was the attack with tanks at Cambrai in November 1917, when a penetration of five miles was secured, but lost a few days later owing to lack of reserves with which to follow up this success. Premature use of the surprise new weapon, plus insistence by the General Staff that tanks should be used in penny packets instead of en masse, soon ruined any chance of these early armoured fighting vehicles becoming a decisive factor.

The Germans evolved some highly successful methods of tactical warfare during the later stages of the war. By establishing thin outpost lines of pillboxes and machine-gun nests, with reserves held ready in the rear to counter-attack the enemy as soon as he got beyond his protective barrage, they avoided the easy targets presented to enemy artillery by heavily-manned front-line trenches. In 1918, Ludendorff devised an alternative to the suicidal waves of attacking infantry by developing a method of infiltration in which German stormtroopers moved forward in groups. First they probed the enemy line, discovered weak points and penetrated them, and then went far ahead fast, leaving stubbornly defended points to be mopped up by the reserves coming up behind.

The battles against the Russians on the Eastern front hold little promise for the wargamer campaigning with model soldiers, consisting as they did mainly of huge and ill-managed Russian armies being defeated by the better strategy and tactics of the Germans. But there are other 'side shows' which can be studied and reproduced on the table-top terrains, such as that part of the war between the Austrians and the Italians in the Dolomites, and the invasion of Gallipoli, in which an amphibious operation was partially successful in spite of heavy losses caused by the Turks' pre-knowledge of the attack. The bitter fighting that took place on that peninsula involved such renowned soldiers as the Australian infantry and their tough enemy, the Turk. Turkey also provided the opposition in Mesopotamia where, after a disastrous start, spectacular victories were won which had had no real influence on the course of the war. As in the Palestine campaign, cavalry were used here, a more open style of warfare which better lends itself to the wargames table. In Palestine, there was a similar type of warfare, with the added spice of Lawrence and his Arabs ambushing armoured trains and conducting guerrilla warfare behind the enemy lines. The operations against German and native forces in South-East Africa can only be described as a 'side show to the side shows'. Nevertheless, perhaps these long-drawn-out campaigns fought over vast expanses of the African continent offer more to the wargamer than any other aspects of World War I. Here we find a mixture of European and native infantry, cavalry, ramshackle motor vehicles alongside horse transport, some primitive armoured cars and engagements on rivers and lakes between gunboats dragged for hundreds of miles across country and painfully put together by the lakeside. In addition, a German cruiser, trapped up a long African creek, used its crew and guns as a shore party to assist in the campaign.

Though it is, of course, impossible to consider military history and the development of tactics without reference to World War I, there are nevertheless, few wargamers who consider the period to be interesting enough to reproduce on the table-top. Perhaps they are wrong, and perhaps what has been written here may stimulate some campaigners with model soldiers to explore a new and so far untapped field of operations.

How does one go about reflecting such a vast, complex and sprawling conflict on to the limited confines of the wargames table? It has already been mentioned that certain aspects of World War I do lend themselves to battles with model soldiers and it is on these that we will concentrate, ignoring the large-scale onslaughts where masses of men rose from one trench system and attempted to battle across a stretch of no-man's-land tightly barred by wire entanglements and under murderous fire from massed machine-guns and artillery.

The open warfare of 1914 seems merely to be an extension of 'horse-and-musket' warfare, plus the added involvement of machine-guns together with more numerous and longer-range artillery. Cavalry still attempted to perform their traditional role on the battlefield and infantry still fought with their personal weapons in the shelter of natural obstacles and out in the open. The machine-gun, although not at this period in such numbers as in later days of the war, naturally caused heavier casualties than one normally gets in 'horse-and-musket' wargames. This must be taken into account when formulating rules for the period, and a simple method which reasonably simulates the arc of fire of the machine-gun is as follows.

Each machine-gun is allocated three dice per move of firing, with results scored in the same way as though one were throwing three dice for three volleys of infantry musketry fire. To reproduce the traversing effect of the machine-gun, allow the gun to cover a frontage of say 18 in and use one dice for each of the three 6-in sections of that distance. For example, in the first third of the move the machine-gun is allowed to fire on a 6-in zone in its right, deducting from the dice or scoring for casualties in whatever way the rules allow. In the second third of the move, the machine-gun traverses and fires on the 6-in zone directly to its front, whilst in the last third of the move the machine-gun traverses over a 6-in zone to its left, scoring in the same way. Of course, the machine-gun may traverse from left to right if preferred, or it may concentrate all its fire in one zone.

A practice peculiar to World War I was the trench raid, in which a small party of infantry, faces blackened and armed with close-contact weapons such as daggers and clubs, crept across no-man's-land and raided the opposite trenches. This was either done to secure a prisoner from whom perhaps seemingly vital information might be extracted or else, more frequently, it was a calculated attempt to maintain an aggressive spirit in the breasts of the front-line troops whilst at the same time keeping the enemy on their toes and denying them any rest. Whatever the reason, trench raids were costly ventures, bringing little tangible gain for the lives that were lost. A trench raid can realistically be represented on the table-top in a fairly simple and inexpensive manner. Ideally, it could take place on a sand-table in which trench systems could be dug and a littered no-man's-land laid out between them. But few wargamers possess a sand-table and an alternative, possibly preferable, is to lay out two Bellona dioramas of the 1914–18 trench system. This is a realistic set-piece with two lines of inter-communicating trenches, complete with firing platforms and revetted sides, with board, corrugated iron and sandbags. Two of these pieces laid one on each side of the table with a suitably littered no-man's-land between them can form an interesting battleground to fight over. The figures required will be any of the World War I participants—British infantry, German infantry, French infantry or American infantry. The defending side garrisons its trench as required, whilst the attackers plan the manner in which they are going to get across without being detected, surge down into the enemy trenches, grab their prisoner and get back with him to their own lines.

It has already been stressed that surprise and concealment are the most difficult factors to reproduce in battles with model soldiers, and as these are the very essence of this 'trench raid' game, consideration must be given to means of simulating it. If the raiding party are laid out in full view of the commander of the opposing side he will immediately see all that is going on and move his defending troops accordingly. Therefore it must be decided whether the attacking troops are actually set up on the table in their progress across no-man's-land, or whether their journey is marked on some map or diagram and they are not revealed until they actually come charging down into the enemy trenches. An extremely simple method—ensuring more than realistic confusion, since no one can see what they are do-ing—is to have the wargames table illuminated only by a small 5-watt lamp painted blue! In this gloomy light, the little men will have to move across their battlefield and carry out their raid whilst the enemy move to meet it. The effects of Verey lights or star shells can be simulated by the momentary flash of a torch aimed upwards at the ceiling and reflecting its light downwards. This method is more amusing than practical but it does, at least, convey something of the dark confusion of the real thing.

The alternative is to plot the movement of the trench-raiding party across no-man's-land, giving the defending general the chance to detect them. Because the game is laid on and some-thing has to happen, he *knows* that a raid is coming over whereas in real life it might not have occurred until the next night or the night after that. Although perhaps elaborate in its conception and execution, the following method has a reasonable chance of working and of realistically reproducing the surprise element so essential to a trench raid.

Lay out the terrain in the manner suggested and then draw a map in duplicate of the entire area, scaled to perhaps 2 in or 3 in to the foot. This means that if the table on which the battle is taking place is, say, 4 ft square, the map will be 12 in × 12 in with the trench systems and barbed wire entanglements, shell holes, etc, drawn in to the same scale. Of course, with a battle-ground of that size more than one section per side of the Bellona trench system will be required and, ideally, three or four sections

laid alongside each other will be needed on each half of the
battlefield. Each commander has a piece of tracing paper or
transparent talc or plastic which he lays over his map. The game
is now ready to begin and each commander marks on his piece
of talc the exact dispositions of *each man* in his force at the
start of the game. Having carried out their first moves, each
commander again marks any movements of his individual men.
The movement of each man will be to scale, ie, if the move
allowed on the table is 12 in then on the map the man will move
3 in. During each move the defender will have the opportunity
to use some of the methods of detection which have been decided
prior to the start of the game: star shells, machine-guns firing,
and the rifle shots of sentries. These will be reflected on the
wargames table with the essential aid of a third man, or umpire.
At the beginning of each move, the defending commander will
tell the umpire privately whether he intends using any of these
defensive methods during the forthcoming move.

If he decides to use a star shell or a Verey light, the umpire
will freeze the move halfway through and, by studying the
attacker's map, ascertain whether any of the attackers were
standing up and would have been exposed to view by the star
shell, or whether they were under any form of cover and so in-
visible to the defenders. The umpire will then mark on the
defender's map the locations of the men who were visible but
the attacker will not be given this information. The defender
will now have the chance to fire at the exposed men providing
his garrison are sufficiently alert: so he throws a dice and requires
a 5 or 6 to fire whilst the man being fired upon has the chance of
throwing a 5 or 6 to save himself if he receives that fire.

If the defender is utilising one of his two available bursts of
machine-gun fire, he then marks the path of the bullets by a
dotted line on his map (preferably in a different colour). When
the umpire receives the two maps at the end of the move and
compares them he will be able to see if the fire has hit any of the
advancing attackers who will then be required to throw a dice to
see if they were killed or only wounded. A similar method can
be used to reproduce the effect of rifle shots by sentries.

When the attackers reach the trench system (on the map) their actual figures are placed on the battlefield. Similarly, the figures of the defender's garrison are placed in position in their trench system as they are marked on their map. The attackers now tumble into the trench and a mêlée ensues, using whatever rules one desires to simulate this hand-to-hand conflict.

To secure a prisoner, the attackers single out one, two or three men (however many they wish to try and capture) and two men are detailed to escort each prisoner back across no-man's-land. Before getting the prisoner out of the trench, the attackers will need to throw a dice and score 4, 5 or 6.

Once out of the trench, the attackers move back visibly for one move and then, together with their prisoners, are removed from the table and their progress again takes place on their map under the same conditions as their advance. If any succeed in returning to their own trenches accompanied by a prisoner, the attackers are then considered to have won the game, otherwise victory goes to the defenders. The umpire must consider, at the moment when the attackers tumble down into the enemy trenches, just how much of a surprise their attack has turned out to be. If it is a complete surprise then he is entitled to award them two points to be added to each dice they use in the mêlée (presuming that the mêlée will be conducted on an individual basis of man against man, highest dice winning). If it is a partial surprise, he allows them to add one point per dice.

Another factor which will add to the interest and realism of the game is to have occasional shells arriving explosively on the scene without the control or knowledge of either commander. This will be under the control of the umpire who will have, say, two British and two German shells at his disposal. His map should be marked off in numbered squares so that he can tell by throwing a dice or drawing a playing card from a pack exactly where the shells will land. Thus he may decide during the third move to have a German shell land somewhere near the British trenches, or a British shell may land in or near the German trenches. Having decided where it lands, he then informs the commander who might be affected and that commander has to throw a dice for any men within a 2-in or 3-in radius of the point at which the shell burst.

Another small-scale engagement typical of World War I could also be fought on a terrain resembling that used for a trench raid, where two opposing lines of trenches face each other separated by shell-pocked and littered no-man's-land. Across this area will lurch a tank or tanks of the primitive type used by the British in 1917. The tanks are highly suspect so far as mechanical failures are concerned, so that at the start of every move they will need to throw a dice and, if a 1 comes up, then the tank has broken down. At the start of the next move another dice is thrown and, providing it does not throw a 1 or 2, the tank is free to move again. In the event of a 1 or 2 coming up, then a 1, 2 or 3 will halt the tank at the start of the next move, and so on. The Germans, lacking adequate anti-tank guns, used field-guns firing over open sights and some deadly high-calibre anti-tank rifles. Certain Germans can be nominated as possessing these weapons, while others can be given the faculty of clambering on the tank and putting a grenade through a gun slit or porthole. The tanks can be handicapped by getting bogged down or lurching into a shell hole too large for them to emerge from but, should they arrive unscathed at the enemy trenches, their effect will be devasting. If the trenches are wide, an additional refinement can be included by having the tanks carrying large rolled bundles of wood (fascines) which drop down to fill the trench and allow the tank to cross.

The landings at Gallipoli were amphibious operations and differed from those on the Normandy beaches in 1944 in that, in 1915, there were no landing-craft which could beach directly above water level. At Gallipoli, large ocean-going vessels such as the *River Clyde* beached themselves and lowered specially-made gangways, down which the troops poured under heavy fire from the prepared Turkish positions. This operation lends itself to wargaming primarily as part of a campaign of sorts in which a map of the coastline is drawn and the defender is then allowed to allocate his troops to that area whilst the attacker can select the point on the coast at which he will attempt to land. Of course, neither commander knows the dispositions of the enemy, ie, where they are strong on the ground in defensive positions or at what point the landing is to be attempted. This means that the defender might well have to rush troops from another sector during the actual battle, and it also gives the attacker the opportunity to make a feint attack at one point while putting in his real effort at another.

The actual fighting and firing will be much the same as 'horse-and-musket' with the added fire-power of machine-guns. An additional realistic accompaniment to the battle will be what is known as 'off the table' firing from the heavy guns of naval vessels lying out to sea. These can be allocated to the attacker at so many shells per move to fall in numbered areas of the battle-field decided by the throw of the dice or the drawing of a card. In other words, the attacker announces that a battleship is firing a 15-in shell and then decides (by dice or card) its point of landing. Should there be troops in the vicinity, he then works out how many of them are killed within a radius of so many inches. The size of the killing radius obviously depends upon the size of the shell, more men are likely to be killed by a big shell than by one of a smaller calibre.

The land operations in Mesopotamia, Palestine and German East Africa all lend themselves very well to table-top battles with model soldiers. Really, they are 'horse-and-musket' games with the addition of armoured cars, machine-guns, armoured trains, irregular native soldiers, and native troops with both British and German forces. In Mesopotamia, cavalry played a vital role and there is considerable scope for wide-scale cavalry operations, ideally planned on a map as part of a campaign. In German East Africa, the wargamer has scope for mixing early wheeled mechanical vehicles, horse-drawn vehicles, native troops (Askaris with the Germans, and Indian and other coloured troops with the British). Engagements can take place in open country or under semi-jungle conditions and, if a campaign is attempted, one can involve landing parties from ships or use the guns of the ships to aid in land battles (using the 'off-table' firing system already explained).

15
World War II

If WORLD WAR I lacks its wargaming devotees then World War II certainly makes up for the deficiency. It is not unreasonable to claim that between a third and a half of today's table-top generals find fascination in the reproduction of its many complex aspects: a multitude of new weapons, both heavy and light; armoured vehicles of widely differing sizes and strengths; air power, ranging from the infantry-supporting dive-bomber and fighter to the heavy bomber and its effect upon the capabilities of a nation to wage war; sea warfare and amphibious operations, and so on. Add to these the vast number of soldiers of many nations fighting all over the world under conditions ranging from the intense cold of the Russian winter to the heat of the Pacific islands and the Libyan desert, and the infinite potentialities for the wargamer are self-evident.

The general pattern of World War II is so well known and its history so profusely documented that it will suffice here to recommend the campaigner wishing to reproduce some of its many facets to make a special study of the following:

German blitzkrieg tactics, using armour and dive-bomber support in Poland, the Low Countries and France. Wavell's crushing defeat of the Italian army in Libya.

The 'colonial-type' campaigns against the Italians in Somaliland and Abyssinia.

The German paratroop invasion of Crete.

The German invasion of Russia.

Eighth Army versus Afrika Corps—the various campaigns in the Libyan desert culminating in El Alamein and the North African invasion.

The commando raids on enemy-occupied Europe.

The Japanese successes at Hong Kong and Singapore.

The vast tank battles and land operations in Russia.

The Allied campaign in Tunisia.

The capture of Sicily and the invasion of Italy.

The British retreat in Burma and subsequent reconquest of that country.

The Russian successes in Eastern Europe.

The Allied campaign in Italy—Cassino, Anzio and the battles for the Gothic Line.

D-Day and the Normandy invasion.

The battle for Normandy and the breakthrough to Paris.

The invasion of Germany and battles for the Rhineland.

The airborne operations at Arnheim.

The American amphibious operations against the Japanese in the Pacific.

This list is probably more notable for its omissions than for what it includes. No mention has been made of the vast air operations of the war, such as the Battle of Britain and the daylight bombing raids of Germany. Naval wargamers may shake their heads sadly at the omission of such vital operations as the Battle of the River Plate or the sinking of the *Bismarck*. The answer, of course, is that innumerable books have been written on all these facets of the war, and only by studying the full details of such operations will the man who campaigns with model soldiers discover the necessary information to reproduce them in miniature.

Since World War II, warfare has become so complex and embraces such a wide variety of weapons that any consideration of the problem of transferring it to the wargames table is far beyond the scope of this book. Indeed, it could be said that a wargamer could spend the greater part of his wargaming time in studying the potentialities and effects of modern weapons and equipment and in formulating sets of rules which would never be complete and with which he would never be satisfied. So, in the belief that it is better to get down to actual battles on the table-top than to ponder endlessly on the complex ramifications of modern warfare, the following basic rules are offered for World War II-type battles with model soldiers.

Move Distances

Infantry	6 in	+ 3 in on road
Jeep	18 in	+ 3 in on road
15-cwt; 3-ton trucks	12 in	+ 3 in on road
Armoured cars	15 in	+ 3 in on road
Bren carriers; ⎱ Scout cars; APCs ⎰	18 in	+ 3 in on road
Quad and 25-pdr ⎱ 88-mm and truck ⎰	12 in	+ 3 in on road
Self-propelled guns	12 in	
Churchill tank	10 in	
Panther tank; Sherman ⎱ tank and German ⎬ assault guns ⎰	12 in	
Tiger tank	8 in	

Firing

Pistol	6 in	(6 reqd @ 6 in; 5/6 @ 3 in)
Grenade	6 in	(1 dice per grenade—kill half score)
Sub-machine gun	9 in	do s/mg do
Rifle	15 in	(Throw 1 dice per 3 men)—@ 15 in–3
Bren gun	15 in	(Throw 2 dice per gun)—@ 9 in–2
		—@ 6 in–1
*Panzerfaust ⎱ *Bazooka ⎰	6 in	4, 5 or 6 to score hit Attack value=4
Flame-thrower	6 in	2 dice each—kill half score. Enemy within 6′ in of men flamed, throw for morale—½ run 12 in otherwise stand
*Pak a/tk gun	24 in	Attack value = 2. Initial dice for hit
*6-pdr a/tk gun	36 in	do = 3. do
*75-mm gun	36 in	do = 4. do 3 in b/circle
*88-mm gun	36 in	do = 5. do 3 in do
*25-pdr gun	36 in	do = 5. do 3 in do
*Mortar	9 in–24 in	Use 3– in burst circle
Hy m/gns	18 in	3 dice per gun—@ 18 in–3 ⎱ @ 12 in–2 ⎬ per dice @ 6 in–1 ⎰

* Initial dice throw for accuracy—At 36 in need 6   ⎱
  (Deduct 2 from dice if target   24 in need 5 or 6   ⎬ to score
   is moving)   12 in need 4, 5, or 6   ⎬ hit
      6 in need 3, 4, 5 or 6   ⎰

Troops behind hard cover—only *half* casualties.

Some points of explanation are needed in connection with the firing section of these rules. Primarily they concern the 'attack-value', which is a system of reflecting the effects of heavy weapons upon mainly armoured vehicles. Each weapon is given an attack-value corresponding to its potentialities when its calibre, weight of projectile and penetrative power are taken into consideration. It can be seen that the 6-pdr anti-tank gun, for example, has a considerably lower attack-value than the fearsome and highly effective 88-mm gun used by the Germans. Similarly, it will be seen when dealing with the potentialities of armoured fighting vehicles that they have an attack-value for their armament and a defence-value relative to their speed and armour-thickness. To utilise these attack- and defence-values, let us quote as an example a 6-pdr anti-tank gun firing upon a German Panther tank.

The attack-value of 6-pdr anti-tank gun is 3, while the defence-value of a Panther tank is 12. However, if the missile strikes the tank on its thinner, armoured side, then 1 is deducted from the defence-value, similarly 2 is deducted when the hit is on the even less-armoured rear. Conversely, if the tank is hull-down (in a protected position behind rising ground or a wall etc, so that only its turret is showing) 1 is *added* to the vehicle's defence-value. Let us suppose that the 6-pdr is firing at a range of 24 in, which necessitates a dice score of 5 or 6 to register a hit. Allowing that this condition has been satisfied and a hit registered on the front plates of the the Panther, then the 6-pdr anti-tank gun is required to equal or beat the tank's defence-value of 12. Let us assume that the anti-tank gunner throws two dice which total 9—added to the guns attack-value of 3, this gives a total of 12. This means that the Panther tank has been knocked out. Had the score been less than 12, then the tank would have been safe.

Another necessary clarification of the firing section concerns the 'burst-circle'. This is indication of the explosive power of a projectile and the circle is large or small in proportion to the size of that projectile. For example, it can be seen that a mortar (which can drop its bomb in an area stretching from 9 in in front of the mortar itself to 24 in) has a burst-circle of 3-in radius.

This circle can be made by a ring of wire or a cut-out of transparent plastic. The centre of the circle is placed exactly over the bursting point of the shell or bomb, and anything covered by the circle or within its bounds is considered to have been hit. Obviously, if the projectile is bigger than that of a mortar then the burst-circle will be correspondingly larger.

We have already discussed the attack- and defence-values of weapons and armoured vehicles. What follows is a simple set of tables outlining the capabilities, both offensive and defensive, of the principal British and German tanks and anti-tank guns.

Churchill Tank

Move	10 in
Attack-value	= 3
Defence-value	= 10
Attack on side of tank	deduct 1
Attack on rear of tank	deduct 2
Hull-down	add 1
Against personnel	4-in b/circle
	2 m/gs

If two hits are scored on a tank without knocking it out, it remains stationary for the rest of the game but may fire.

Tiger Tank

Move	8 in
Attack-value	= 5
Defence-value	= 12
A/Personnel	4- in b/c, 2 m/gns
Attack on side	deduct 1
Attack on rear	deduct 2
Hull-down	add 1

If two hits are scored on tank without knocking it out, it remains stationary for the rest of the game but may fire.

Self-Propelled Guns

Move	12 in
Attack-value	= 4
Defence-value	= 8
Attack on side	deduct 1
Attack on rear	deduct 2
Hull-down	add 1
A/Personnel	4-in burst circle

If two hits are scored on this vehicle without knocking it out, it remains stationary but may fire.

88-mm Gun and Vehicle

Move	12 in
Attack-value	= 5
Defence-value on move or hidden	= 8
Defence-value stationary	= 4
A/Personnel	4-in burst circle

Takes 3 in off move to limber or unlimber.

Two hits without knocking the vehicle out will make it immobile. Gun is automatically out of action on second hit.

Sherman Tank

Move	12 in
Attack-value	= 4 (17-pdr = 5)
Defence-value	= 9
Attack on side	deduct 1
Attack on rear	deduct 2
Hull-down	add 1
A/Personnel	4-in burst circle, 2 m/gns

If two hits are scored on the tank without knocking it out, it remains stationary for the rest of the game but may fire.

Panther Tank

Move	12 in
Attack-value	= 5
Defence-value	= 12
Attack on side	deduct 1
Attack on rear	deduct 2
Hull-down	add 1

If two hits are scored without knocking the tank out, it remains stationary for the rest of the game but may fire.

A/Personnel   2 m/gns and 4-in burst circle

Bren Carrier and 6-pdr Anti-Tank Gun

Move	18 in
Attack-value	= 3
Defence-value on move or hidden	= 8
do stationary	= 4

Two hits without knocking out the carrier will render it immobile for the rest of the game.

Takes 3 in of move to limber or unlimber.

25-pdr Gun and Vehicle

Move	12 in
Attack-value	= 5
Defence-value on move or hidden	= 8
do stationary	= 4
A/Personnel	4-in burst circle

Two hits without knocking quad out will render it immobile. Gun is automatically out after two hits.

Takes 3 in of move to limber or unlimber.

Mêlées

In modern warfare, there is relatively little hand-to-hand fighting because one side or the other will be forced by the effect of the enemy's fire to break and run before actually coming to grips. The following simple method gives a reasonable simulation of this aspect of modern warfare:

1. When attackers come within 6 in of defenders, both throw dice and multiply number of troops by dice score. If attackers win, then defenders move back 12 in.
2. If defenders win, attackers cannot move in until they have diced:

    1, 2, 3 = remain stationary for that move.
    4, 5, 6 = move in and mêlée as usual.
3. At any stage of attack (if they move first, for example) defenders may become counter/attackers and move forward, then carry on as in 1.

All these rules are, of course, only basic suggestions designed to meet the wargamer's needs at the outset of his attempts to recreate World War II on the table-top. Rules of a more advanced nature are available from a number of different sources and almost every experienced wargamer has his own set!

When first fighting modern wargames on the table-top it is very easy to overload one's forces with armoured fighting vehicles, such as tanks, self-propelled guns and armoured cars. Unless your rules have detailed and realistic conditions which allow the infantry to fight back with anti-tank grenades or with PIATs and bazookas, they will stand no chance of success, the tank will become mistress of the battlefield and the only result will be a tank *v* tank contest. So use tanks sparingly, give perhaps a hundred infantrymen a supporting troop of three tanks and make sure that the infantrymen are well supplied with anti-tank guns. Only in this way will a realistic battle be possible on the table-top.

The long range of the heavy weapons of World War II was such that it is not only impossible but would be completely unrealistic to place artillery on the table-top. To do so would mean that, on the scale of the area over which one is fighting, the guns would be firing over open sights at a few hundred yards range whereas in reality they would be dropping their shells from four or five miles behind the actual fighting. For this reason it is desirable that all artillery should be 'off the table' and should fire in the 'burst-circle' manner already described, or by some similar method.

Air power, as we know, played a very decisive part in World War II and it will be recalled that the rout of the Allied armies in France and the Lowlands during 1940 was largely due to the support given to the German infantry by the dive-bomber. It is, however, extremely difficult to simulate air power in a table-top wargame and, again, it may be necessary for attacking aircraft to be 'off the table'. However, if the wargamer is particularly keen to use aircraft made up from many of the excellent kits on the market, methods of doing so have been evolved and are to be found in specialist books. (*See* Bibliography.)

16
Wars of Today

SINCE THE end of World War II, there have been only two relatively major conflicts affording opportunity to test new tactics and weapons. Overshadowing all military operations hangs the threatening cloud of the nuclear weapon—the 'ultimate deterrent' which can render every other weapon valueless and return men to fighting with sticks and stones. Conscious of this fact, a few wargamers have formulated rules to simulate the use of atomic weapons—one such prophet of doom has a rule which enables his entire wargames table, with the exception of a 6 in margin round its edges, to be cleared at one fell swoop by the use of a nuclear weapon!

The war in Korea was a bitterly-contested modern conflict fought over an uninviting terrain by a predominance of American troops backed by others of British, Commonwealth, Turkish and other nationalities. Initially, their opponents were the North Koreans, who were later aided by the Chinese. Apart from the fact that it presents an interesting opportunity to operate with the forces of many countries, the Korean war did not introduce any particularly new tactics, or any weapons markedly different from those used in World War II. However, it might appeal to the wargamer as a modern conflict fought with infantry, armour and air power, but lacking the complications and/or refinements of the greater conflict which preceded it.

The prevailing conflict in Vietnam between the Americans and the Viet Cong is an ominous example of how a small and comparitively backward nation, armed and equipped by a major power, can by using guerrilla methods which take every possible advantage of local terrain and conditions, contain the superior numbers and vastly more sophisticated equipment of a far more powerful nation. Often unable to distinguish by their appearance between friend and foe and constantly confused by the sudden transformation of a ruthless soldier seeking to kill him into an apparently harmless peasant, small wonder that the American soldier has found this to be a frustratingly hard conflict.

Apart from the loosely organised fighting methods of the Viet Cong, this conflict's special interest to the wargamer might well lie in the unprecedentedly large-scale use of helicopters to transport soldiers speedily from one point of battle to another, together with the modern practice of conveying infantry as far forward as possible in armoured personnel carriers. This employment of airborne infantry, in particular, is a modern practice which can usefully be transferred to the wargames table. Other factors worthy of consideration are the modern infantryman's personal weapon—a repeating carbine which enables an immensely heavy rate of fire-power to be maintained—recoiless artillery weapons, and the highly effective armour-piercing projectiles currently in use.

The war in Korea can be fought along much the same lines as those used for World War II. It will be necessary for the wargamer to work out attack and defence values for Russian equipment being used by the Chinese and North Koreans, and it may be necessary for him to consider differing morale values for the various troops involved in this war.

So far as table-top reproduction of conflicts in Vietnam is concerned, the representation of American troops, and simulation of their way of fighting and their weapons, presents no great problem. It is when one considers the enemy, soldiers one minute and peasants the next, that one comes up against immense difficulties in formulating rules which will realistically represent their activities. Emerging behind American lines from long tunnels, or suddenly transformed from peasants again into soldiers, their organisation, like their uniforms, is so nebulous as to make them an uninteresting type of model soldier to collect. It is possible that this war is only one of a number which the Western world will be forced to wage against Communism and in which the enemy will be much of a kind with the Viet Cong and fighting under the same banner. If this type of war holds any appeal for the man who battles with model soldiers, he will probably find most interest in small-scale actions in which sections or platoons of infantry move forward in armoured personnel carriers such as the 'General Sheridan', or arrive from the skies in a vertical assault by helicopter. It is also more than likely that, in the near future, vertical take-off and landing aircraft will supplant the slower and more vulnerable helicopters.

This sort of warfare lends itself to a high degree of preliminary map-moving by contestants and may tend to develop into a situation where the lighter-armed Viet Cong can only be brought to bay by the more heavily-armed Americans when they are left with no alternative. The resulting battle will probably be one in which the morale of the opposing sides will have to be closely considered, particularly when the overwhelming American fire-power seeks out the elusive and hidden enemy—all of which is exceedingly difficult to simulate on a wargames table.

Not all battles with model soldiers, however, seek to reproduce actual conflicts and the wargamer is free to use his imagination, to ignore major powers with nuclear forces more or less counteracting each other, and to invent 'fringe' wars fought with conventional weapons by relatively small numbers of fighting men who are as professional as the English longbowman or the Roman legionary. This is the day of smaller, more cohesive groups of heavily-armed professionals, masters of their trade using weapons of double or treble the fire-power of those used a decade ago. The simulation of these weapons by means of wargames rules is, however, far from easy and there is the danger that either both sides will be wiped out early in the game, or else blanket-fire will be laid on all points that might harbour the enemy, so that luck takes over from skill.

The personal weapons of the modern infantryman are also far more versatile than before. His rifle is capable of being used in a semi-automatic or fully automatic role, and he can be armed with shoulder-fired grenade launchers, recoiless rifles, or wire-guided anti-tank missiles with inbuilt self-correcting devices that keep it lined up on target. Artillery, too, is lighter, shoots further and uses ammunition with greater fragmentation, larger bursting charges and more powerful propellants. Anti-tank projectiles now make life exceedingly unpleasant for the armoured soldier because the penetrative power of high-explosive anti-tank (HEAT), high-explosive squash head (HESH), and armour-piercing-discarding Sabot (APDS), makes them truly formidable weapons.

All of this is of sufficiently recent occurrence to necessitate considerable research so that wargame rules may be involved incorporating appropriately scaled-down ranges and effects which will give a realistic simulation of modern warfare. The biggest problem is likely to be the size of the wargames table that will be required if the scaled-down ranges are to be anything like accurate!

APPENDICES

Appendix 1
Author's Note

READERS OF this book will, I think, fall into clearly defined categories. A large proportion of boys and men, whilst not necessarily wishing to don uniform and endanger their own lives whilst seeking that of an enemy, possess a strong sense of tradition and a pride in much of British military history. Such people, attracted by the book's title and dust-jacket, may well pick it from the shelves of a bookshop and, in the modern jargon, make an 'impulse' purchase.

Then there is another, most likely a boy, who has recently bought, or had bought for him, a box or two of those very attractive plastic model soldiers and who feels that there ought to be something more sensible to do with them than just knock them down with marbles rolled across the floor or table. How right he is, and if he follows this instinct he will find that he has opened a door to one of the most exciting, fascinating and stimulating of all hobbies or pastimes.

Or perhaps a friend of a friend has a fine collection of model soldiers and is known to reconstruct battles like Waterloo and Gettysburg with them, fighting for hours on end over miniature scenic terrain until he has reversed military decisions in a manner that could have changed the face of the world had it occurred at the time. So the friend is sought out and he is introduced to the owner of the soldiers who, delighted to encounter someone genuinely interested, sets about initiating him into the mysteries of battles with model soldiers. Both will be saved many hours of discussion, explanation and demonstration if the novice has previously read a book such as this. Not only does it put him into the picture but, even more important, it will stimulate him to gather together armies of his own with which to wage war against his mentor.

Mark the word 'stimulate', because that is what this book is all about. Nothing in these pages is a dictate, no word says you must or you shall do it this way. On the contrary, the book sets out from the very beginning to stimulate the reader to think for himself, and to use what he has read merely as a foundation for efforts and ideas which reflect his own temperament and character. Only in this way will he obtain maximum satisfaction from the hobby of battling with model soldiers.

Wargaming is a many-featured activity which can be everything or anything to all men. Unlike chess, draughts or even football, all games with traditional and time-proven sets of rules, battles with model soldiers are fought to gloriously elastic rulings born of their inventor's conception of what a cavalry charge was like, or the effect of grapeshot against a Napoleonic square.

General Sherman, of American Civil War fame, is quoted as saying, 'War is hell.' So it is, and perhaps the wargamer, seeing just how helpless his little plastic figures are against the dice-simulated effects of cannon and muskets, will appreciate more than ever the utter futility of real war.

DONALD FEATHERSTONE

Appendix 2
Converting Airfix Figures

BECAUSE THIS is a book for beginners and because the inexpensive range of Airfix figures is particularly well-suited to the beginner, it has been frequently mentioned in this text. Extensive and steadily-increasing as the Airfix range is, many types of soldiers mentioned in our survey of military history are at present unobtainable, but there is no soldier of any period, type or equipment that cannot be converted from existing production figures. This opens wide the door not only to battles with model soldiers in any period, but also to a fascinating and practical pursuit that will bring pride to the hobbyist, conscious that his figures are different from any others in the world! Conversion is not difficult, simple methods readily suggest themselves and are often a matter of plain common sense. However, expert advice is available, as explained below.

*Airfix Magazine** regularly publishes articles describing the conversion of Airfix figures into those of other types and periods. Well illustrated and simply written by modellers of experience and ability, these articles are invaluable to both the novice and the practiced wargamer. Each includes comprehensive instructions, with details of the uniforms, weapons and equipment of the specific period, together with a useful historical background.

Up to the time of writing, the following articles had been published, and may still be obtainable in back numbers of the

* Published on the fourth Friday on each month and available from bookstalls and hobby shops, or by subscription from Surridge Dawson & Co Ltd, 136–42 New Kent Road, London SE1.

issues in which they appeared. Failing this, they should be begged or borrowed for they are essential literature.

'Modelling the 8th Army', by A. V. Martin. (December 1964)

'Modelling the Red Army', by C. O. Ellis. (April 1965)

'Making Japanese Infantry Equipment', by C. O. Ellis. (April 1965)

'British Uniforms and Equipment', by C. O. Ellis. (August 1966)

'My Crimean War', by Colin Jones. (April, May, June 1967)

'The Zulu War', by Colin Jones. (July, August, September 1967)

'The US Civil War', by Michael Blake. (October 1967–February 1968)

'The German Army 1914–1918', by David Nash. (March to October 1968)

'Conversion of Airfix Romans into Allies and Enemies', by Bob O'Brien. (February 1968 and November 1968 to June 1969)

'The British Army '14–18', by David Nash. (July 1969 onwards)

Other magazines, such as *Slingshot* of the Society of Ancients,* regularly publish articles on conversions. All are worthy of study.

Appendix 3
Suppliers of Model Soldiers
Suitable for Wargames

Airfix Products Ltd (toyshops, model shops, woodwork, etc).

Douglas Miniatures, J. Johnston, 17 Jubilee Drive, Glenfield, Leicester, LE3 8LJ.

The Garrison, Castlegate, Knaresborough (Nr Harrogate), Yorks.

Les Higgins, 52 High Street, Hardingstone, Northampton.

F. Hinchcliffe, 83 Wessendenhead Road, Meltham, Yorks (guns only).

Hinton Hunt Figures, Rowsley, River Road, Taplow, Bucks.

W. H. Lamming, 130 Wexford Avenue, Greatfield, Kingston-upon-Hull, Yorks.

Miniature Figurines, 5 Northam Road, Southampton, or 100A St Mary's Street, Southampton.

Oscar Figures—Harry Novey, Longcroft, The Green, Little Horwood, Bletchley, Bucks.

Rose Miniatures, 45 Sundorne Road, Charlton, London SE7.

Tradition, 188 Piccadilly, London W1.

Willie Figures, 60 Lower Sloane Street, London SW3.

Command Post, 760 West Emerson, Upland, California 91786, USA.

Hobby House, 9622 Ft Meade Road, Laurel, Maryland 20810, USA.

C. H. Johnson, PO Box 281, Asbury Park, New Jersey 07713, USA.

Jack Scruby, Box 89, 2044 South Linwood Avenue, Visalia, Calif 93277, USA.

K. and L. Company (late Thomas Industries), 1929 North Beard, Shawnee, Okla 74801, USA.

SPAIN

Alymers Miniploms, Maestro Lope 7, Burjasot, Spain.

SWEDEN

Holgar Eriksson, Sommarrovagen 8, Karlstad, Sweden.

WEST GERMANY

Elastolin (O. and M. Hausser) Neustradt/Coburg, Jahres-wende, West Germany.

Flat Figures

Aloys Ochel, Feldstrasse 24b, Kiel, West Germany.

Rudolf Donath, Schliessfath 18, Simback/Inn OBB, West Germany.

F. Neckel, Goethestrasse 16. Wendlingen Am Neckar, West Germany.

Appendix 4
Wargames Clubs

At the time of writing, the following list contained all the known wargames clubs and groups scattered throughout Great Britain. It is possible that others have formed since, and it is certain that the name of *your* newly-formed club would add lustre to the list!

BIRMINGHAM Society of Wargamers in. Miss Vivienne Crabb, 'Mariemont', CBCE, Westbourne Road, Edgbaston, Birmingham 15.

BRISTOL Wargame Society. M. Blake, 102 Cotham Brow, Redland, Bristol 6.

CARDIFF Wargame Society. Bill Cainan, 5 St Mark's Avenue, Cardiff, CF4 3NW.

CHELTENHAM Wargames Club. Chris Beaumont, Hampton House, Shurdington Road, Cheltenham, Glos.

DORKING Kriegspiel Society. Stuart Etherington, 6 Dell Close, Mickleham, Nr Dorking, Surrey.

DUNDEE Wargaming Club. John G. Robertson, Upper Dunglass, Arbroath Road, West Ferry, Dundee.

EASTBOURNE Wargames Society. David Spencer, 70 Langney Rise, Eastbourne, Sussex.

ESSEX AND SUFFOLK Wargames Club. Geoffrey Kelker, Camelot, St Jude Close, Parsons Heath, Colchester, Essex.

HARLOW Wargame Club. D. O. Watson, 125 Hookfield, Harlow, Essex.

HORNCHURCH Wargame Club. Henry Trevillyan, 96 North Street, Hornchurch, Essex.

IRISH Model Soldier Group. Ian Green, 14 Gledswood Avenue, Clonskeath, Dublin 14, Eire.

KNARESBOROUGH Wargame Club. John Turnbull, 43 Woodpark Drive, Knaresborough, Yorks.

LEEDS Wargame and Military History Society. C. A. Sapherson, 55 Wensley Drive, Leeds 7, Yorkshire.

LEICESTER Wargame Club. T. J. Halsall, 8 Westleigh Road, Leicester.

LONDON Wargame Section. The Secretary, 90 Burrage Road, Plumstead, London SE 18.

MANCHESTER University Wargames Society. Mr Ley, c/o Moberly Tower Hall of Residence, Burlington Street, Manchester.

MERSEYSIDE Wargamers Club. N. J. Hunter, 8 Egremont Prom, Wallasey, Cheshire.

MID-HERTS Wargames Group. H. Gerry, 39a Sandpit Lane, St Albans, Herts.

MIDHURST (Sussex) Wargames Group. Chris Speedy, 3 Elmleigh, Midhurst, Sussex.

NAVAL WARGAMES Society. The Secretary, 16 Hugo Road, London N19.

NORTH EASTERN Wargaming Club. John G. Robertson, Upper Dunglass, Arbroath Road, West Ferry, Dundee.

NORTH LONDON Wargames Group. The Secretary, 14 Midhurst Avenue, Muswell Hill, London N10.

OSWESTRY Wargames Club. P. Davis, Gable End, 1 Oakhurst Avenue, Oswestry, Salop.

PLYMOUTH SOUND Naval Wargame Association. B. J. Carter, 56 Purbrook Gardens, Purbrook, Portsmouth, Hants.

RICHMOND Wargame Club. Stewart Gordon-Douglass, 4 Enmore Court, East Sheen, London SW14.

SEVERN VALLEY Wargame Section. J. E. Hammond. 10 Richmond Road, Cardiff, South Wales.

SOUTH SUSSEX Wargames Club. Malcolm Woolgar, 44 Shaftesbury Avenue, Worthing, Sussex.

SOUTH GLOUCESTERSHIRE Wargames Society. The Secretary, 24 Grange Park, Frenchay, Nr Bristol.

THANET Wargames Society. The Secretary, 16 Princes Gardens, Cliftonville, Margate Kent.

TUNBRIDGE WELLS Wargame Society. George Gush, 154D Upper Grosvenor Road, Tunbridge Wells.

WALSALL Wargames Club. E. Stubbs, 22 Willow Road, Great Barr, Birmingham 22A.

WALTHAM FOREST Wargame Club. The Secretary, 12 Warner Road, Walthamstow, London E17.

WESSEX Military Society. D. F. Featherstone, 69 Hill Lane, Southampton, SO1 5AD, Hants.

WORCESTERSHIRE Warriors Association. James Opie, Flat 2, Laurel Garth, 45 Graham Road, Malvern.

Bibliography

THE BEST-KNOWN books on wargaming include the following:
Little Wars, H. G. Wells. (1913)
Wargames, Donald Featherstone. (1962)
Wargames in Miniature, J. Morschauser. (1963)
Naval Wargames, Donald Featherstone. (1966)
Air Wargames, Donald Featherstone. (1967)
Charge!, Brigadier P. Young and Lt-Col J. P. Lawford. (1967)
Advanced Wargames, Donald Featherstone. (1969)
Discovering Wargaming, John Tunstill. (1969)
Introduction to Battle Gaming, Terence Wise. (1969)
Wargames Campaigns, Donald Featherstone. (1970)
Most are of fairly recent origin and readily available—even
Wells's classic (the 'Bible' of wargaming) has been reprinted in
an excellent photo-copied edition and can be obtained through
Wargamers Newsletter, the magazine of the hobby, at 69 Hill
Lane, Southampton, SO1 5AD, England.
There follows a selection of literature which will aid the war-
gamer in assembling his table-top armies.
Making and Collecting Military Miniatures, Bob Bard. (1959)
Model Soldiers: A Collector's Guide, John G. Garratt. (1960)
Lead Soldiers and Figurines, M. Baldet. (1961)
Model Soldiers, H. Harris. (1962)
Model Soldier's Guide, C. Risley and W. Imrie. (1964)
Tackle Model Soldiers This Way, D. F. Featherstone. (1965)
Collecting Toy Soldiers, Jean Nicollier. (1967)
Model Soldiers, R. B. R. Nicholson. (1967)
How to go Model Soldier Collecting, H. Harris. (1969)
Handbook for Model Soldier Collectors, D. F. Featherstone.
(1969)

Military Modelling, D. F. Featherstone. (1970)

A History of the Regiments and Uniforms of the British Army,
  R. M. Barnes. (1950)

British Military Uniforms, W. Y. Carman. (1957)

European Military Uniforms, Paul Martin. (1963)

Military Uniforms and Weapons, R. Toman. (1964)

Soldiers, Soldiers, R. Bowood. (1965)

Military Uniforms of the World, P. Kannick. (1968)

Dress Regulations for the Army (reprinted 1970)

Le Costume et les Armes des Soldats de tous les temps (Vols I
  and II), Fred and Liliane Funcken. (1968)

L'Uniforme et les Armes des Soldats du Premier Empire, Fred
  and Liliane Funcken. (1968)

Cavalry Uniforms of Britain and the Commonwealth, Robert
  and Christopher Wilkinson-Latham. (1969)

French Army Regiments and Uniforms, W. A. Thorburn.
  (1969)

Besides discussing the background to wargaming and collecting armies for table-top battles, this book also describes soldiers and their tactics throughout the ages. There are a vast number of reference books to these subjects, many of which are now out of print and very difficult to find. The following selection of recently published books will, however, be found of assistance and may also encourage further reading of the books listed in their respective bibliographies. Most of them can easily be obtained from bookshops or libraries, and many have been published in cheap paper-backed editions—these are marked*.

Decisive Battles of the Western World, J. F. C. Fuller. (1956)

Battles that Changed History, Fletcher Pratt. (1956)*

Fighting Men, Treece and Oakeshott. (1963)

Hastings to Culloden, Young and Adair. (1964)

From Spearman to Minuteman, S. E. Ellacott. (1965)

Conscripts on the March, S. E. Ellacott. (1965)

Weapons and Tactics: Hastings to Berlin, Jac Weller. (1966)

The Fighting Man, Jack Coggins. (1966)

The Ancient Art of Warfare, J. Bourdet. (1969)

Great Battles of Biblical History, Sir R. Gale. (1968)

206

Battles with Model Soldiers

The Art of Warfare in Biblical Lands, Y. Yadin. (ND)
Alexander of Macedon, Harold Lamb. (1955)*
The Greek and Macedonian Art of War, F. E. Adcock. (1957)
The Generalship of Alexander the Great, Col J. F. C. Fuller. (1958)
Hannibal, H. Lamb. (1958)*
Enemy of Rome, L. Cottrell. (1960)*
The Roman Legions, H. M. D. Parker. (1928)
The Roman Art of War Under the Republic, F. E. Adcock. (1960)
Seeing Roman Britain, L. Cottrell. (1967)*
The Great Invasion, L. Cottrell. (1958)*
The Roman Imperial Army, Graham Webster. (1969)
The Art of War in the Middle Ages AD 378–1515, C. W. C. Oman. (1953)
Anglo-Saxon Military Institutions, C. W. Hollister. (1968)
The Mongols, E. D. Phillips. (1969)
The Military Organization of Norman England, C. Warren Hollister. (1968)
A History of the Crusades, Steven Runciman. (1951–4)
In the Steps of the Crusaders, R. Pernoud. (1959)
The Crusaders, Henry Treece. (1964)*
Sieges of the Middle Ages, Philip Warner. (1968)
The Crecy War, Lt-Col Burne. (1955)
The Agincourt War—A Military History of the Latter Part of the Hundred Years War from 1369–1453, Lt-Col A. N. Burne. (1956)
Agincourt, Christopher Hibbert. (1964)
The Sword and the Age of Chivalry, R. E. Oakeshott. (1964)
A Knight and His Armour, R. E. Oakeshott. (1965)
A Knight and His Horse, R. E. Oakeshott. (1965)
A Knight and his Weapons, R. E. Oakeshott. (1965)
The Bowmen of England, Donald Featherstone. (1967)
A History of the Art of War in the 16th Century, Charles Oman. (1937)
Army Royal—Henry VIII's Invasion of France 1513, C. G. Cruickshank. (1963)

Elizabeth's Army, C. G. Cruickshank. (1946)

Charles XII of Sweden, R. M. Hatton. (1968)

The King's Peace 1637–1641, C. V. Wedgewood. (1955)

The Thirty Years War, C. V. Wedgewood. (1957)*

The Great Civil War (1642–1646), Lt-Col A. H. Burne and Lt-Col P. Young. (1959)

Battles of the English Civil War, Austin Woolrych. (1961)*

Battles and Generals of the Civil Wars, Col H. C. B. Rogers. (1968)

Civil War in England, J. Lindsay. (1968)

Edgehill, Brigadier P. Young. (1966)

The Civil War, Richard Atkyns. (1967)

Oliver Cromwell and His Times, Brigadier P. Young. (1962)

Cromwell's Army, C. H. Firth. (1921)*

The Regimental History of Cromwell's Army, C. H. Firth and G. Davies. (1940)

Duke of Marlborough, His Life and Times, W. S. Churchill. (1934–8)

*Marlborough and His Campaigns 1702-1709**, Lt-Col A. Kearsey. (1960)

The Marlborough War, Robert Parker and Comte de Merode-Westerloo. (1968)

The Captain-General—The Career of John Churchill, Duke of Marlborough 1702–1711, Ivor F. Barton. (1969)

Prince Eugene of Savoy, N. Henderson. (1965)

The Armies of Queen Anne, Major R. E. Scouller. (1966)

England Under Queen Anne, G. M. Trevelyan. (1946)

Field-Marshal Lord Ligonier—A Story of the British Army 1702–1770, R. Whitworth. (1958)

Battles of the '45, K. Tomasson and F. Buist. (1967)*

Culloden, J. Prebble. (1961)*

His Brittanic Majesty's Army in Germany During the Seven Years War, Sir Reginald Savory. (1966)

General Wolfe at Quebec, C. Hibbert. (1959)

The Capture of Quebec, Christopher Lloyd. (1959)

The Plains of Abraham, B. Connell. (1960)

The Battle of Plassey, M. Edwardes. (1963)

208

The Siege of Gibraltar 1779–1783. T. H. McGuffie. (1965)
The Campaigns of Napoleon, D. Chandler. (1967)
The Anatomy of Glory (*Napoleon's Imperial Guard*), H. Lachouque and A. Brown. (1961)
Wellington in the Peninsula, Jac Weller. (1962)
Waterloo, John Naylor. (1960)*
Wellington at Waterloo, Jac Weller. (1967)
Victorian Military Campaigns, Ed Brian Bond. (1967)
At Them With the Bayonet!, D. F. Featherstone. (1968)
All for a Shilling a Day, D. F. Featherstone. (1966)
North West Frontier, Arthur Swinson. (1967)*
The Sound of Fury (*Indian Mutiny*), Richard Collier. (1963)*
The Foreign Legion, Charles Mercer. (1964)*
The Destruction of Lord Raglan, C. Hibbert. (1961)*
The Battles of the Crimean War, W. Baring Pemberton. (1962)*
The Reason Why, Cecil Woodham Smith. (1953)*
Gettysburg, Miers and Brown. (1962)*
Arms and Equipment of the Civil War, J. Coggins. (1962)
Ordeal by Fire, Fletcher Pratt. (1950)*
The Guns at Gettysburg, Fairfax Downey. (1958)*
The 20th Maine, John J. Pullen. (1959)
The Franco-Prussian War, Michael Howard. (1961)
Goodbye Dolly Gray (*Boer War*), Rayne Kruger. (1959)*
Washing of the Spears, D. Morris. (1966)*
The River War, W. S. Churchill. (1960)*
Frontiers and Wars, W. S. Churchill. (1962)
The Siege of Peking, P. Fleming. (1962)*
The Guns of August (*1914*), Barbara Tuchman. (1964)*
The Vanished Army, Tim Carew. (1964)
Gallipoli, Alan Moorhead. (1964)*
The Somme, A. H. Farrer-Hockley. (1964)*
In Flanders' Fields, Leon Woolf. (1961)*
The Donkeys, Alan Clark. (1961)*
1918, Barry Pitt. (1966)*
The World War 1939–1945, Brigadier P. Young. (1966)

The War 1939–1945, Ed Desmond Flower and James Reeves.
  (1960)
The Struggle for Europe, C. Wilmot. (1951)*
The Sands of Dunkirk, R. Collier. (1961)*
The Desert Rats, G. L. Verney. (1954)*
Alamein and the Desert War, Ed D. Jewell. (1967)*
The Desert Generals, C. Barnett. (1960)*
The Battle for North Africa, J. Strawson. (1969)
The Italian Campaign, G. A. Sheppard. (1969)
The Battle for Rome, W. G. F. Jackson. (1969)
The Battle for Normandy, E. Belfield and H. Essame. (1965)*
The Battle for Germany, H. Essame. (1969)
Russia at War, A. Werth. (1965)*
Barbarossa, Alan Clark. (1960)*
Pork Chop Hill (Korea), S. L. A. Marshall. (1956)*
This Kind of War (Korea), T. R. Fehrenbach. (1963)

Index